Optics and Applications

Optics and Applications

Dr. Lazo M. Manojlović

ARCLER
PRESS

www.arclerpress.com

Optics and Applications

Dr. Lazo M. Manojlović

Arcler Press
2010 Winston Park Drive,
2nd Floor
Oakville, ON L6H 5R7
Canada
www.arclerpress.com
Tel: 001-289-291-7705
001-905-616-2116
Fax: 001-289-291-7601
Email: orders@arclereducation.com

ISBN: 978-1-77361-222-5 (Hardcover)

Library of Congress Cataloging-in-Publication Data

A CIP record for this book has been applied for and is available from Library and Archives Canada.

ABOUT THE AUTHOR

Dr. Lazo M. Manojlović

Dr. Lazo M. Manojlović was born in Vranje, Serbia, on April 30, 1972. He received the B.Sc.E.E. and M.Sc.E.E. degrees in electrical engineering from the Faculty of Electrical Engineering, University of Belgrade, Belgrade, Serbia, in 1996 and 2003, respectively, and the Ph.D. degree in electrical engineering from the Faculty of Technical Sciences, University of Novi Sad, Novi Sad, Serbia, in 2010.

From 1997 to 1998, he was with the Institute Mihailo Pupin, Belgrade, Serbia, as a part of his extended studies, where he was engaged in research on the development of applicative software for text to speech synthesis of Serbian spoken language in the C programming language based on diphone concatenation. From 1998 to 2002, he was with the Military Institute of Technology, Belgrade, Serbia, where he was engaged in research on designing and testing several types of ultralownoise, wideband photodiode preamplifiers for quadrant photodetector and for tracking electronics circuits for analog signal processing in pulsed-laser tracking systems for laser guidance missiles. From 2002 and 2003, he was with Pupin Telecom DATACOM, Belgrade, Serbia, where he was involved in designing, installing, and commissioning of the first hybrid fiber coaxial network for the cable distribution systems in Belgrade. From 2003 to 2005, he was with the Vienna University of Technology, Vienna, Austria, where he was engaged in research on the development of high-precision piezoactuator- based positioning systems. From 2005 and 2008, he was with Integrated Microsystems Austria GmbH, Wiener Neustadt, Austria, where he was engaged in research on the development of different types of fiber-optic sensors. He is currently with the

Ministry of Telecommunications and Information Society of the Republic of Serbia, Belgrade, Serbia, as a Special Adviser to the Minister, and also with the Zrenjanin Technical College, Zrenjanin, Serbia, as a Professor. His research interests include interferometry, optical sensing systems, and laser rangefinders.

TABLE OF CONTENTS

LIST OF ABBREVIATIONS

AC	Alternating Current
APD	Avalanche Photodiode
BER	Bit Error Rate
CATV	Common-Antenna Television
CCD	Charge-Coupled Device
CIE	Commission Internationale de l'Eclairage
CIPM	Comité International des Poids et Mesures
CW	Continuous-Wave
DC	Direct Current
DWDM	Dense Wavelength Division Multiplexing
FDDI	Fiber Distributed Data Interface
FWHM	Full Width at Half Maximum
HDTV	High-Definition Television
IR	Infrared
LAN	Local Area Network
LCS	Low-Coherence Source
LED	Light Emitting Diode
MAN	Metropolitan Area Networks
NA	Numerical Aperture
PLL	Phase-Locked Loop
PMD	Polarization Mode Dispersion
PMT	Photomultiplier
SNR	Signal-to-Noise Ratio
TE	Transverse Electric

TM	Transverse Magnetic
WDM	Wavelength Division Multiplexed
WLS	White-Light Source
YAG	Yttrium Aluminum Garnet

LIST OF FIGURES

LIST OF TABLES

PREFACE

Optics and Application deals with the applicative aspects of optics and optical phenomena with its main emphasis on applying the optical knowledge onto solving the real engineering problems. The book is aimed toward undergraduate and graduate students who want to broaden their knowledge of optics to the real life engineering problems as well as to the engineers who want to acquire the basic principles of optics.

The book is arranged in a way which leads the reader toward better understanding of optical phenomena and their applications. Basic tools and concepts are presented in the earlier chapters, which are then developed more and applied in the later chapters. The book is organized in nine chapters covering broad range of optical phenomena.

Chapter 1 gives a brief historical overview of optical phenomena perception. Although the light and light related phenomena have always been a source of immense curiosity for ancient peoples, one had to wait until the late eighteenth century and the Maxwell's classical electromagnetic theory in order to conceive most principles of modern optics

Chapter 2 introduces the radiometry as the substantial part in the field of optical measurement and engineering together with its measurement techniques for measuring optical power and its spectral content.

Chapter 3 describes the light propagation in terms of rays within the geometrical optic framework of treating the optical phenomena. The geometrical optics represents a cornerstone for designing many optical instruments such as telescopes, microscopes, etc.

Contrary to the Chapter 3, where light propagates in a form of a simple beam, Chapter 4 takes into account the wave nature of light thus introducing the corresponding interference phenomena. Being inherently short wavelength electromagnetic wave the interference of light gives a rise of highly sensitive measurements of many physical quantities, which are also presented in this chapter.

Chapter 5 further develops the wave based phenomena of light such as diffraction.

Chapter 6 introduces the holography and the practice of making holograms. A hologram is a photographic recording of a light field, rather than of an image formed by a lens. The hologram is used to display a fully three-dimensional image of the holographed object.

Today's state-of-the-art communication systems are unimaginable without optical fibers. Therefore, Chapter 7 is devoted to the optical fibers and their use in the communication as well as in the highly sensitive versatile measurement systems.

Chapter 8 deals with the polarization phenomena of light and introduces a simple mathematical formalism in order to determine the polarization state of light.

Last but not least Chapter 9 shows how the light can be detected. Although in the past times human eye was used exclusively as an optical detector, in order to objectively measure the intensity of a light in modern optical systems a solid-state detector has been usually used.

Author

SYNOPSYS

Optical phenomena and their applications draw attention since the revival of human civilization. Being familiar with the rectilinear propagation of the rays of light ancient Greeks managed to perform exceptional astronomical measurements of their time only by using simple interplay between the light and shadow. Aristarchus of Samos (c. 310 BC – c. 230 BC) measured the Moon diameter, the Moon to Earth distance, as well as the distance between the Earth and Sun only by using geometrical relations between the illuminated and shadowed part of the Moon. Eratosthenes of Cyrene (c. 276 BC – c. 195/194 BC) calculated for the first time the Earth's circumference by simple measurement of the geometrical parameters of the corresponding shadows into two wells placed at local noon on the summer solstice in Syene (modern Aswan, Egypt) and in Alexandria. Knowing the distance between Syena and Alexandria and the measured Sun's angle of elevation at noon on the same day in Alexandria the calculation of Earth's circumference was straightforward. The next quantum leap in the application of optical phenomena was reached by the discovery of curved optical surfaces and their reflective and refractive phenomena. The first optical instruments based on the curved optical surfaces were telescopes used for magnification of distant objects, and microscopes used for magnifying very small objects. Since the days of their inventions by Galileo and Van Leeuwenhoek, these optical instruments have been greatly improved. Today's application of optics and corresponding optical phenomena ranges from the state-of-the-art measurement systems to the state-of-the-art communications systems.

1 CHAPTER

THE NATURE OF LIGHT

THE BRIEF HISTORY OF OPTICAL PHENOMENA PERCEPTION

Since the dawn of human civilization the man was astonished and puzzled by the everyday interplay between the day – light and the night – dark caused by the Sun. Being unable to understand the origins of observed optical phenomena but understanding the importance of such phenomena on everyday life, the early men ascribed to the Sun the divine imprint. For a long period of human history the Sun, the Moon, and the stars where the only sources of light which periodic temporal behavior were well understood and interweaved into the human perception of time. Sometimes this perfectly tempered clock was interrupted by the solar and Moon eclipses as well as by the lightning and the polar light that confused the early man perception of light. The first turning point in understanding and controlling the optical phenomena and being able for the first time to make a man controlled light source was the discovery of fire control. For the first time the primordial man was able not only to heat his home but also to illuminate it.

The curious mind of the ancient Greek philosophers was the first one who tries to get a deeper insight in the nature of light. The early Greek philosopher Pythagoras (c. 582 BC – c. 497 BC) believed that light, which originates from the visible objects, is carried by the tiny particles of light. Empedocles (5th century BC) believed that light came from the illuminating objects but also believed that the light rays came out from our eyes. He was the first one who postulated that light travels at a finite speed. The famous Greek mathematician Euclid (c. 325 BC – c. 265 BC) thought that the eyes transmit rays of light that cause the sensation of vision. Mirrors were also studied by Euclid, where in his book, entitled *Catoptrics* and written by him in the 3rd century BC, the law of reflection was established.

The rectilinear propagation of rays of light was the main reason for Isaac Newton in his *Treatise on Opticks* to adopt the corpuscular nature of light, more than twenty centuries after Euclid. He believed that light consists of very small bodies emanating from the shining bodies. The shadow formation behind the illuminated objects as well as the law of mirror reflections goes into the favor of corpuscular nature of light. In contrast to Newton's corpuscular theory, Christian Huygens developed different theory where light is considered to be a wave motion that spreads from the source uniformly in all directions. The optical phenomena such as interference and diffraction, which were inexplicable in the frame of Newton's corpuscular theory, perfectly fit to the wave nature of light.

In the mid 19th century one of the milestones in understanding the nature of light was reached. Namely, due to the genius work of James Clerk Maxwell that was published in his famous paper *A Dynamical Theory of the Electromagnetic Field* the light can be described as an electromagnetic wave. However, the beginning of 20th century brings new dilemmas. The outstanding experimental results, firstly announced by German physicist Philipp Lenard in 1902, showed that the energy of electrons emitted from the illuminated surface increased with the frequency of the light. The classical electromagnetic theory crashes in trying to explain this phenomenon. Therefore, a new approach was necessary. This new approach came on the wings of new quantum theory where light has been considered to consist of discrete wave packets – photons.

The presented brief history of optical phenomena perception showed that throughout the entire history of human preoccupation with the light there was back and forth movements between the wave and corpuscular nature of light. The Maxwell's classical electromagnetic theory perfectly describes the propagation of light whereas the quantum theory treats the interaction between light and matter or the absorption and emission of light. So, if someone ask "*What is light?*" there is no a simple answer to this simple questions.

LIGHT AS AN ELECTROMAGNETIC WAVE

Maxwell's equations together with the Lorentz force law represent the complete classical theory of the electromagnetic fields. All the phenomena where the interaction between the electromagnetic field and matter occur can be successfully explained within the frame of classical electromagnetic theory except in those cases where the quantum effects are distinct. The set of Maxwell's equations can be written in the following form:

$$\nabla \cdot \mathbf{E} = \frac{\rho}{\varepsilon_0}, \qquad (1.1)$$

$$\nabla \cdot \mathbf{B} = 0, \qquad (1.2)$$

$$\nabla \times \mathbf{E} = -\frac{\partial \mathbf{B}}{\partial t}, \qquad (1.3)$$

$$\nabla \times \mathbf{B} = \mu_0 \left(\mathbf{J} + \varepsilon_0 \frac{\partial \mathbf{E}}{\partial t} \right), \qquad (1.4)$$

where **E** is the electric field vector (or electric field strength), **B** is the magnetic induction, ρ is the electric charge density, **J** is the electric current density, $\varepsilon_0 = 8.85\times10\text{-}12$ F/m is the vacuum permittivity, and $\mu_0 = 4\pi\times10\text{-}7$ H/m is the vacuum permeability. The force f per unit volume acting on the electric charge density ρ and the electric current density J is given by the Lorentz force law:

$$\mathbf{f} = \rho\mathbf{E} + \mathbf{J} \times \mathbf{B}. \qquad (1.5)$$

If the electromagnetic field is observed in the regions where are no electric charges ($\rho = 0$) and no electric currents ($\mathbf{J} = 0$), such as in a vacuum, Maxwell's equations reduce to:

$$\nabla \cdot \mathbf{E} = 0, \tag{1.6}$$

$$\nabla \cdot \mathbf{B} = 0, \tag{1.7}$$

$$\nabla \times \mathbf{E} = -\frac{\partial \mathbf{B}}{\partial t}, \tag{1.8}$$

$$\nabla \times \mathbf{B} = \varepsilon_0 \mu_0 \frac{\partial \mathbf{E}}{\partial t}. \tag{1.9}$$

By taking curl ($\nabla \times$) of the curl equations and by using the identity $\nabla \times (\nabla \times \mathbf{X}) = \nabla(\nabla \cdot \mathbf{X}) - \nabla^2 \mathbf{X}$ one can obtain the following wave equations:

$$\nabla^2 \mathbf{E} = \frac{1}{c^2} \frac{\partial^2 \mathbf{E}}{\partial^2 t}, \tag{1.10}$$

$$\nabla^2 \mathbf{B} = \frac{1}{c^2} \frac{\partial^2 \mathbf{B}}{\partial^2 t}, \tag{1.11}$$

where c is the speed of light in vacuum identified as:

$$c = \frac{1}{\sqrt{\mu_0 \varepsilon_0}} = 299{,}792{,}458 \, \frac{\mathrm{m}}{\mathrm{s}}. \tag{1.12}$$

Although the speed of light in vacuum has the exact value, the speed of light v in matter depends on the medium characteristics such as relative permittivity ε_r and relative permeability μ_r in the following way:

$$v = \frac{1}{\sqrt{\mu_r \mu_0 \varepsilon_r \varepsilon_0}} = \frac{c}{n}, \tag{1.13}$$

where $n = \sqrt{\mu_r \varepsilon_r}$ is the material index of refraction. The electric field **E** and magnetic induction **B** of an electromagnetic wave are mutually perpendicular to each other and to the direction of wave propagation and are in phase with each other, i.e. the wave propagate in the direction of $\mathbf{E} \times \mathbf{B}$ as it is presented in figure 1.1.

If we take into consideration the Cartesian coordinate system and resolve the vector wave equations (1.10) and (1.11) into components, one can observe that both components **E** and **B** satisfy the general scalar wave equation:

$$\frac{\partial^2 U}{\partial x^2}+\frac{\partial^2 U}{\partial y^2}+\frac{\partial^2 U}{\partial z^2}=\frac{1}{v^2}\frac{\partial^2 U}{\partial^2 t}, \qquad (1.14)$$

where U represents any one of the electric field components E_x, E_y, and E_z, as well as any one of the magnetic induction components B_x, B_y, and B_z. Let further consider the simplest case, where the spatial variation of U occurs only along some particular direction, for example along the z axis, i.e. $U=U(z, t)$. In this particular case equation (1.14) simplifies to one dimensional wave equation:

$$\frac{\partial^2 U}{\partial z^2}=\frac{1}{v^2}\frac{\partial^2 U}{\partial^2 t}. \qquad (1.15)$$

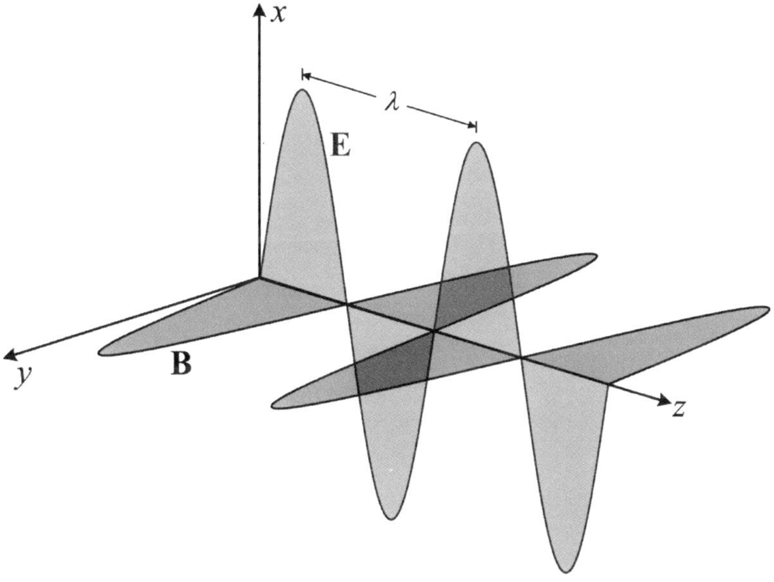

Figure 1.1. The electric field **E** and magnetic induction **B** are mutually perpendicular to each other and to the direction of wave propagation and are in phase with each other.

A simple harmonic solution of this one dimensional wave equation can be obtained by the variable separation in the following form:

$$U(z,t)=U_0\cos(\omega t-k), \qquad (1.16)$$

where U_0 is the corresponding field amplitude and where wave number k and angular frequency ω are related as:

$$v = \frac{\omega}{k}, \quad (1.17)$$

where the ratio of the angular frequency ω and the wave number k represents the phase velocity v. The presented particular solution, given by equation (1.16), is fundamental to the study of optics and represents the plane harmonic wave. The wavelength λ is defined as the distance, measured along the direction of the wave propagation, where the wave function completes one whole cycle, i.e. the distance between two maximums (or minimums) of the wave function. The wavelength is determined by the wave number as:

$$\lambda = \frac{2\pi}{k}. \quad (1.18)$$

If we make a step back to the three–dimensional wave equation (1.14) it is easy to show that the similar simple harmonic solution is one particular solutions of this wave equation. A three–dimensional plane harmonic wave function is given by:

$$U(x, y, z, t) = U_0 \cos(\omega t - \mathbf{k} \cdot \mathbf{r}), \quad (1.19)$$

where the position vector **r** is defined as:

$$\mathbf{r} = \hat{\mathbf{i}}x + \hat{\mathbf{j}}y + \hat{\mathbf{k}}z, \quad (1.20)$$

where $\hat{\mathbf{i}}$, $\hat{\mathbf{j}}$, and $\hat{\mathbf{k}}$ are the unit vectors ($|\hat{\mathbf{i}}| = |\hat{\mathbf{j}}| = |\hat{\mathbf{k}}| = 1$) along x, y, and z axis, respectively. The propagation vector or the wave vector **k** can be also given in terms of its corresponding components as:

$$\mathbf{k} = \hat{\mathbf{i}}k_x + \hat{\mathbf{j}}k_y + \hat{\mathbf{k}}k_z, \quad (1.21)$$

where k_x, k_y and k_z are the corresponding wave vector components along x, y, and z axis, respectively. The magnitude of the wave vector is equal to the wave number, which is given in terms of the corresponding wave vector components as:

$$|\mathbf{k}| = k = \sqrt{k_x^2 + k_y^2 + k_z^2}. \quad (1.22)$$

If we further inspect the equation (1.19) and its cosine function argument $\mathbf{k}\cdot\mathbf{r}-\omega t$, one can observe that the constant values of this argument define a set of planes in space named as surfaces of constant phase (equiphase surfaces) given by:

$$\omega t-\mathbf{k}\cdot\mathbf{r}=\omega t-\left(k_x x+k_y y+k_z z\right)=const. \qquad (1.23a)$$

By interpreting the equation (1.23a), it is obvious that wave vector **k** is normal to the surfaces of constant phase as shown in figure 1.2. Moreover, the constant phase surfaces are moving in the direction of the wave vector with the velocity equal to the phase velocity:

$$v=\frac{\omega}{k}=\frac{\omega}{\sqrt{k_x^2+k_y^2+k_z^2}}. \qquad (1.23b)$$

Electromagnetic waves are produced by oscillating electric charged particles, so the frequency of oscillation determines the type of emitted radiation. Such produced electromagnetic disturbance propagates through space as a monochromatic wave, which is characterized by a single frequency, or polychromatic wave, which consists of many frequencies, either discrete or in a continuum. The energetic content of an electromagnetic wave, which is distributed among the various constituent waves, is called the spectrum of the radiation. Typically, to the different regions of the electromagnetic spectrum specific names are given, such as radio waves, micro waves, optical waves (infrared, visible and ultraviolet light), X–rays, gamma rays, and cosmic rays due to the different way of its production or detection. The common names, descriptions as well as typical frequency ranges are presented in figure 1.3. where the electromagnetic spectrum is given in both terms frequency ν and wavelength λ. One can recall that these two quantities, frequency and wavelength, are linked as:

$$\lambda\nu=c. \qquad (1.24)$$

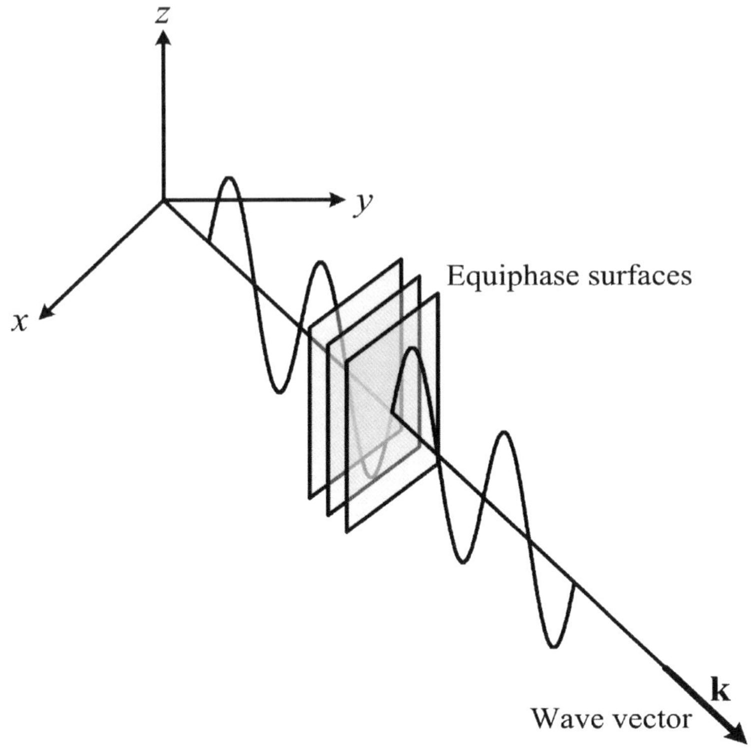

Figure 1.2. Equiphase surfaces of a plane wave.

In the case when all the charges oscillate unison, for the emitted electromagnetic radiation is said to be coherent. In the case when all the charges oscillate independently and/or randomly, for the emitted electromagnetic radiation it is said to be incoherent. Typical radiation sources in the optical domain such as flames, incandescent light bulbs, fluorescent lamps, light emitting diodes (LED), etc. are incoherent light sources. In contrast to these incoherent sources of radiation, typical man made radiation sources in the low–frequency range such as radio waves and microwaves are coherent. The only light source in optical domain that emits coherent radiation is laser.

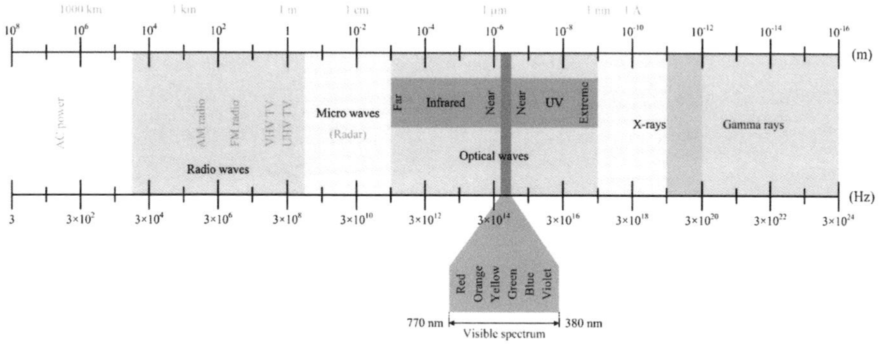

Figure 1.3. Electromagnetic spectrum arranged by wavelength in meters (m) and frequency in hertz (Hz).

Sometimes it is convenient to present the wave function in the complex domain by using Euler representation:

$$U(x,y,z,t) = U_0 \exp[j(\omega t - \mathbf{k}\cdot\mathbf{r})]. \qquad (1.25)$$

The real part of the complex wave function represents the actual physical quantity, where the real part is equal to the actual wave function given by equation (1.19). The reason why the complex representation of the wave function is often used is the simplicity of the algebra with the complex representation than with the trigonometric representation of the wave function. Beside simple harmonic solution of the plain waves that satisfies the wave equation in Cartesian coordinate system, given by equation (1.14), there is also a particular solution in the form of the spherical waves, which can be easily obtained if we rearrange the wave equation given by equations (1.10) and (1.11) in the polar coordinate system. The wave function of the spherical waves is given by:

$$U(r,t) = \frac{U_0}{r}\cos(\omega t - k\ ) \quad \text{or} \quad U(r,t) = \frac{U_0}{r}\exp[j(\omega t - k\ )]. \qquad (1.26)$$

where r is the radial distance from the radiation source to any particular point in the space.

As it was mentioned earlier, the phase velocity is the rate at which the wave phase propagates in space and it is defined as $v_p = \omega/k$. On the other side, the group velocity is defined as the rate at which the wave envelope propagates through space. In order to find the relation for the group velocity we will observe two waves with the angular frequencies ω+dω and ω-dω and wave numbers k+dk and k-dk. In a simple case, which will be considered here, we will assume without losing the generality of the analysis that both waves have the same amplitude U_0 and propagate in the same direction, let say along the z axis. The superposition of both waves will result in a wave which complex wave function is given by:

$$U = U_0 \exp\{j[(\omega + \mathrm{d}\omega)t - (k + \mathrm{d}k)r]\} + U_0 \exp\{j[(\omega - \mathrm{d}\omega)t - (k - \mathrm{d}k)r]\}. \quad (1.27)$$

After the rearrangement of equation (1.27) we have:

$$U = 2U_0 \exp[j(\omega t - k\ )]\cos(t\mathrm{d}\omega - r\mathrm{d}k). \quad (1.28)$$

The last equation can be interpreted as a single wave with the wave function $2U_0 \exp[j(\omega t - k\ )]$ that is modulated by the envelope function $\cos(t\mathrm{d}\omega - r\mathrm{d}k)$. The envelope travels not with the phase velocity but with the group velocity equal to:

$$v_\mathrm{g} = \frac{\mathrm{d}\omega}{\mathrm{d}k}. \quad (1.29)$$

Electromagnetic wave when propagates throughout the space transfers a certain amount of energy and momentum. Therefore, there is a certain electromagnetic energy that is stored in the space through which wave travels. The energy density, stored into the free space is given by:

$$w_E = \frac{\varepsilon_0}{2} E^2, \quad (1.30)$$

$$w_B = \frac{1}{2\mu_0} B^2, \quad (1.31)$$

where w_E and w_B are the energy densities of electric and magnetic fields, respectively. As $E = cB$ is fulfilled, we have $w_E = w_B$. So, the total energy density is given by:

$$w = w_E + w_B = \varepsilon_0 E^2 = \frac{1}{\mu_0} B^2. \quad (1.32)$$

When propagates electromagnetic wave fills the space with the energetic content. If we observe an electromagnetic wave for a very short period of time Δt, the electromagnetic energy that is brought into the observed elementary volume is equal to:

$$W = wAc\Delta t, \quad (1.33)$$

where A is the cross section of the elementary volume. By inspecting equation (1.33) a particular physical quantity can be defined that represents the energy transfer through the unit area and unit time as:

$$S = \frac{W}{A\Delta t} = cw \quad . \qquad (1.34)$$

This physical quantity represents the magnitude of the Poynting vector, which is defined as:

$$\mathbf{S} = \frac{1}{\mu_0}\mathbf{E}\times\mathbf{B} \quad . \qquad (1.35)$$

Besides the energy transfer, electromagnetic radiation carries also the momentum, so when electromagnetic wave impinges the surface of some object it exerts force. As Maxwell showed, the radiation pressure is equal to the energy density of the electromagnetic wave. The radiation pressure exerted on the object that perfectly absorbs the radiation can be found by using virtual work principle. Therefore, we will observe the electromagnetic radiation which impinges the perfectly absorbing body, as it is presented in figure 1.4. If we allow force, which is generated by the radiation, to virtually move the object for an infinitesimally short distance δz, the work done by this force will, according to the first principle, be equal to the change in stored electromagnetic energy. Thus, the energy conservation law gives:

$$PA\,\delta z = wA\,\delta z \quad . \qquad (1.36)$$

where P is the radiation pressure and A is the cross section of the observed elementary volume. From equation (1.36) it is simply obtained $P = w$.

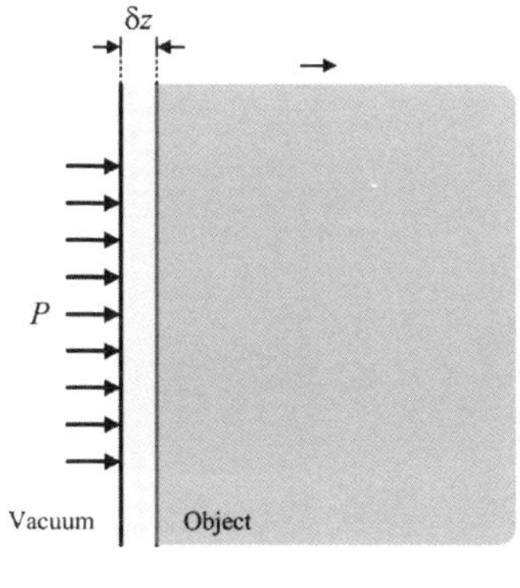

Figure 1.4. Electromagnetic radiation impinges the perfectly absorbing object.

Referring further to figure 1.5., where it is presented the radiation that impinges the perfectly absorbing object, the total momentum of the electromagnetic radiation that is stored in the elementary volume with the cross section A and thickness Δz will be transferred to the object in period of time equal to $\Delta t = \Delta z/c$. The overall momentum M that will be transferred to the object is equal to:

$$M = p_{\mathrm{V}} A \Delta z, \qquad (1.37)$$

where p_{V} is the volume density of electromagnetic momentum. According to the third Newton's law of motion we have:

$$F = M\Delta t. \qquad (1.38)$$

i.e. the force exerted on the object is caused by the overall momentum transfer during the period of time Δt. Therefore, the electromagnetic momentum volume density is given by:

$$p_{\mathrm{V}} = \frac{w}{c} = \frac{S}{c^2}. \qquad (1.39)$$

One of the important physical quantities, especially in the radiometry, is the time average of the energy that is received by the unit area and unit time and is named irradiance. By definition we can write:

$$I = \overline{S}, \qquad (1.40)$$

where $\overline{\bullet}$ denotes the time average operator. So, the irradiance can be interpreted as the time averaged Poynting vector.

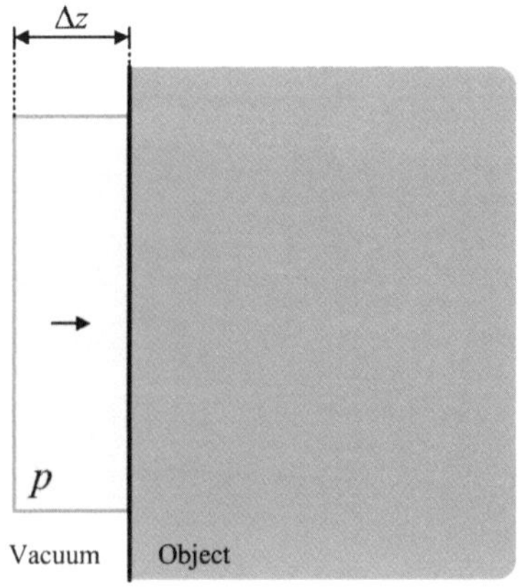

Figure 1.5. Electromagnetic radiation impinges the perfectly absorbing object and transfers the momentum.

If a monochromatic source of electromagnetic radiation is moving uniformly with a constant velocity toward the motionless observer and simultaneously transmits the electromagnetic waves with the particular frequency ν, then, in the observer's frame of reference, according to the relativistic Doppler effect, there will be a change in the received radiation frequency in the following way:

$$\nu' = \nu\sqrt{\frac{c \mp u}{c \pm u}}, \tag{1.41}$$

where ν′ is the received frequency, u is the source velocity relative to the observer, and where upper signs stand for the case when the source is moving away from the observer and lower signs when the source is approaching the observer. Equation (1.41) is obtained by taking into consideration the Lorentz transformation, i.e. equation (1.41) is valid for any velocity. Typically, the source (or the observer) velocity is much smaller than speed of light ($u<<c$), so equation (1.41) simplifies to:

$$\nu' \approx \nu\left(1 \mp \frac{u}{c}\right). \tag{1.42}$$

QUANTUM NATURE OF LIGHT

As mentioned earlier, Lenard's experiment had the far reaching consequences on the deeper understanding of the nature of light. The experiment showed that the energy of emitted electrons is proportional to the frequency of light. The classical electromagnetic theory failed in this case. According to the classical theory, the energy of an emitted electron must be proportional to the intensity of the light but not to its frequency. Moreover, a very dim light would be sufficient to trigger the electron emission that was in opposite to the observed frequency threshold, i.e. the minimal frequency where photoelectric effects starts. Below the threshold, electrons aren't emitted regardless the light intensity or the exposure time. In order to bridge this gap Albert Einstein suggested that a light beam is not a wave that propagates through space, but rather a set of discrete wave packets (photons), each with energy $\varepsilon = h\nu$, with $h = 6.626\times10^{-34}$ Js being the Planck constant. There has been passed almost two decades before the Einstein's light quanta hypothesis had been accepted. The period of transition between the hypothesis rejection and acceptance would undoubtedly have been longer if Compton's experiment had been less striking. Based on the thermodynamic approach to the black–body radiation, Einstein showed that monochromatic black–body radiation in the Wien's law spectral region behaves with respect to thermal phenomena as if it consists of independently moving particles or quanta of radiant energy each with the energy proportional to its frequency.

There is also an alternative way in order to find the relation between the photon energy and the relevant physical quantities of the macroscopic electromagnetic wave. In the analysis shown below, the only assumption that will be taken into account is that electromagnetic radiation is quantized. According to equation (1.19) electromagnetic wave can be represented as:

$$U(\mathbf{r},t) = U_0 \cos(2\pi\nu\, t - \mathbf{k}\cdot\mathbf{r} + \varphi), \qquad (1.43)$$

with φ being an arbitrary chosen initial wave phase. Without loosing the generality of the analysis, the influence of the initial wave phase will be neglected. Therefore, it will be assumed that the photon, as a constitutive part of an electromagnetic wave (radiation), has the energy ε that is dependant on the amplitude U_0, wave vector k, and the frequency ν, i.e. $\varepsilon = \varepsilon(U_0,k,\nu)$. If we go one step back and assume that the initial phase affects photon energy, the photon energy will be dependent on the choice of the initial time, which is in the collision with the energy conservation law and thus absurd. As the wave number is given by $k = 2\varpi/\lambda$ and the wavelength of the electromagnetic wave in vacuum by $\lambda = c/\nu$, the wave number is only dependant on the frequency $k = k(\nu)$. Therefore, for the photon energy we have $\varepsilon = \varepsilon(U_0,\nu)$. Further, in order to find the dependence of the photon energy on the wave amplitude, an isotropic, unvarying, static, and monochromatic source **S** of the electromagnetic radiation will be observed, as it is depicted in figure 1.6.

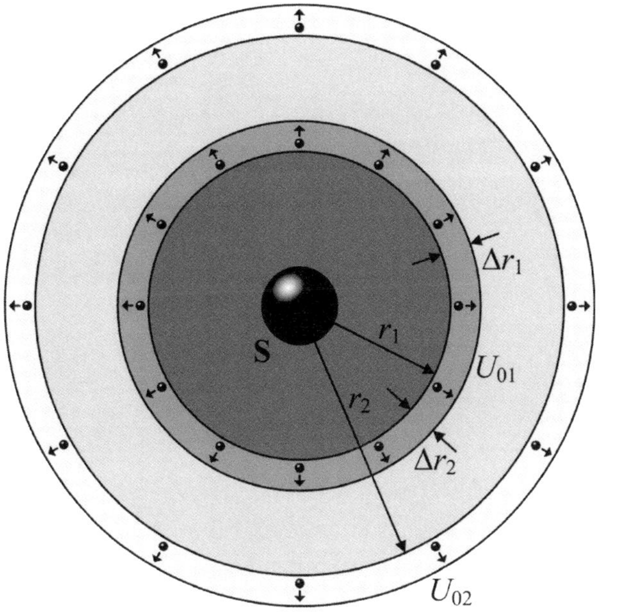

Figure 1.6. Isotropic, unvarying, static, and monochromatic source of the electromagnetic radiation.

At the distance r_1 from the source the amplitude of the radiation is U_{01} and at the distance r_2 the amplitude is U_{02}. In general case as $r_1 \neq r_2$ we have $U_{01} \neq U_{02}$. Since the source is unvarying it emits constant radiant flux Φ, or equivalently it emits a constant number of photons per unit time $\dot{q}$. In the analysis we will consider that the photon velocity is the same as the speed of light. If the consider photon velocity smaller than the speed of light, then some parts of an electromagnetic wave that was firstly broadcasted by the transmitter, will be not quantized at large distances from the transmitter and the others will be, which is absurd. Of course, it is also absurd to consider photons velocity greater than the speed of light as it is in the collision with the basic principle of the special relativity. Therefore, after time $t_1 = r_1/c$ from the moment of their emission, photons reach the distance r_1 from the source. The total number of photons, contained in a thin spherical shell with the thickness Δr_1 at the distance r_1, whereby $\Delta r_1 << r_1$ is satisfied, is $N_1 = \dot{q}\Delta t_1$, where $\Delta t_1 = \Delta r_1/c$, with the assumption that there are no photons absorption and/or scattering on their way since they are traveling in vacuum. Similarly, after time $t_2 = r_2/c$ from the moment of their emission, photons reach the distance r_2 from the source. Similarly, the total number of photons, contained in a thin spherical shell with the thickness Δr_2 at the distance r_2, whereby $r_2 << \Delta r_2$ is satisfied, is $N_2 = \dot{q}\Delta t_2$, where $\Delta t_2 = \Delta r_2/c$, with the same assumption that there are no photons absorption and/or scattering on their way. On the basis of the continuity equation for $\Delta t_1 = \Delta t_2 = \Delta t$ the following is valid $N_1 = N_2 = N$, or due to the invariability of the velocity of photons, all photons contained in the spherical shell defined by the distances $[r_1, r_1+\Delta r]$ after time interval $\Delta t_{12} = (r_2 - r_1)/c$ are contained in the spherical shell defined by the distances $[r_2, r_{2+\Delta}r]$, where $\Delta t = \Delta r/c$ is valid. The energy E_1 of all photons that are contained in the spherical shell at a distance r_1 is $E_1 = N\varepsilon(U_{01},\nu)$, while the energy E_2 of all photons that are contained in the spherical shell at a distance r_2 is $E_2 = N\varepsilon(U_{02},\nu)$. On the first principle basis we have $E_1 = E_2$ that further leads to $\varepsilon(U_{01},\nu) = \varepsilon(U_{02},\nu)$, where it is assumed that there are no external forces that can impact photons motion. As we have fulfilled $\varepsilon(U_{01},\nu) = \varepsilon(U_{02},\nu)$ for any, arbitrarily chosen, amplitudes U_{01} and U_{02}, one can conclude that the amplitude of the electromagnetic wave does not affect the photon energy, so we can finally conclude that the energy of the photons depends only on the frequency of electromagnetic waves, or $\varepsilon = \varepsilon(\nu)$.

Figure 1.7. Transmitter **Tx** emits planar, unvarying, and monochromatic electromagnetic waves in the direction of the receiver **Rx**.

In order to obtain the mathematical relation between the photon energy and the frequency of the electromagnetic wave the transmitter **Tx** that emits planar, unvarying, and monochromatic electromagnetic waves in the direction of the receiver **Rx**, will be considered, as shown in figure 1.7. The transmitter radiant flux and frequency in the transmitter frame of reference are Φ and ν, respectively and are time independent. Since the transmitter flux is constant, the transmitter photon emission rate is also constant and equals $\dot{q}$ in the transmitter frame of reference. These three parameters Φ, $\dot{q}$, and ν of the transmitter are related by the following relation $\Phi = \dot{q}\varepsilon(\nu)$.

The transmitter and receiver antennas are adjusted in such a way that every photon that is being emitted by the transmitting antenna of the transmitter is being received by the receiving antenna of the receiver, i.e. the transmission system has the maximal possible efficiency. If transmitter starts to move away from the receiver uniformly with a constant velocity u, as it is depicted in figure 1.7., due to the Doppler effect the received radiant flux and frequency will differ from the transmitted radiant flux and frequency, where in the receiver frame of reference the received radiant flux and frequency are Φ' and ν›, respectively. Time dilation, caused by the motion of the transmitter in the receiver frame of reference, will change the number of photons per unit time that reaches the receiver to $\dot{q}'$, where in general case $\dot{q}' \neq \dot{q}$ is fulfilled. Following the abovementioned, the average time between the emission of two photons is $\tau = 1/\dot{q}$, while the average time between the capture of two photons is $\tau' = 1/\dot{q}'$. The parameters Φ', $\dot{q}'$, and ν› in the receiver frame of reference are related as $\Phi' = \dot{q}'\varepsilon(\nu')$. According to the Lorentz transformations we have:

$$\nu' = \nu\sqrt{\frac{c-u}{c+u}}, \quad \tau' = \frac{\tau}{\sqrt{1-\frac{u^2}{c^2}}}, \quad \dot{q}' = \dot{q}\sqrt{1-\frac{u^2}{c^2}} \quad . \quad (1.44)$$

Based on the first principle in the receiver frame of reference we have:

$$\Phi' = \Phi - Fu \quad , \quad (1.45)$$

where F is the braking force exerted on the transmitter in the opposite direction of its motion in order to ensure the uniform motion of the transmitter with constant velocity, or the transmitter must not accelerated due to the repulsive force of the electromagnetic radiation that it emits, so we have:

$$\dot{q}'\varepsilon(\nu') = \dot{q}\varepsilon(\nu) - Fu \quad , \quad (1.46)$$

where the last two equations give:

$$\dot{q}\sqrt{1-\frac{u^2}{c^2}}\varepsilon\left(\nu\sqrt{\frac{c-u}{c+u}}\right)=\dot{q}\varepsilon(\nu)-Fu \quad . \qquad (1.47)$$

If the transmitter velocity is close to the speed of light in vacuum, i.e. $u \to c$, equation (1.44) gives $\dot{q}' \to 0$, i.e. the average time between receiving two consecutive photons tends to infinity $\tau' \to +\infty$ and $\varepsilon(\nu') \to \varepsilon(0)$, so regardless the value of $\varepsilon(0)$, the last equation becomes:

$$F=\dot{q}\frac{\varepsilon(\nu)}{c}=\dot{q}p \quad , \qquad (1.48)$$

where p is the photon momentum, and the condition $\varepsilon(0) < +\infty$ has been taken into account, while in opposite the indefinitely large energy will be necessary to establish a time unvarying electromagnetic field even in a limited volume. Based on the third Newton's law, force F is also the repulsive force of the electromagnetic radiation that acts on the transmitter.

The above presented analyze has been performed in order to find the photon momentum as a function of its energy. In the next step, the case when the transmitter is moving steadily towards the receiver with constant velocity u will be considered, as shown in figure 1.8. In order to overcome the repulsive force of the emitted electromagnetic radiation and to ensure uniform motion of the transmitter, the force F must be exerted on the transmitter in the direction of its motion. According to the energy conservation law in the receiver frame of reference the following is valid:

$$\Phi'=\Phi+Fu \quad , \qquad (1.49)$$

By combining equations (1.48) and (1.49) we have:

$$\dot{q}'\varepsilon(\nu')=\dot{q}\varepsilon(\nu)\left(1+\frac{u}{c}\right) . \qquad (1.50)$$

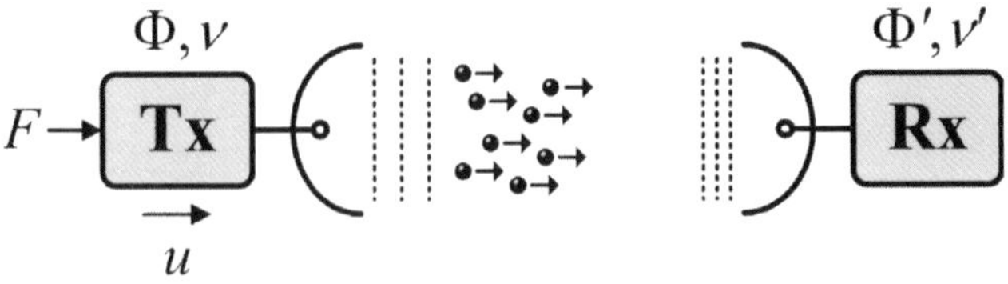

Figure 1.8. The transmitter **Tx** is moving uniformly toward the receiver **Rx** with a constant velocity.

Equations (1.44) and (1.50) give the following functional equation:

$$\varepsilon\left(\nu\sqrt{\frac{c+u}{c-u}}\right)=\sqrt{\frac{c+u}{c-u}}\varepsilon(\nu) \quad . \qquad (1.51)$$

The last functional equation is of the form ε(aν) = aε(ν), where $a=\sqrt{(c+u)/(c-u)}$ and consequently $a \geq 0$. In order to find the solution of this functional equation we will assume that the solution could be found in the following form ε(ν) = νf(ν), where f(ν) is an arbitrary function. Substitution further gives aνf(aν) = aνf(ν) that for the general case ν > 0 further yields to f(aν) = f(ν). Parameter a can take any positive value, so the function f(ν) must have the same, constant value for any argument, or f(ν) = h = *const.* and thus ε(ν) = hν, where h is also the well–known Planck's constant $h = 6.626\times10^{-34}$ Js that was firstly measured by Millikan a century ago. The revealed relation between the photon energy and the radiation frequency confirms the earlier adopted hypothesis ε(0) < +∞, since according to well–known relation we have ε(0) = 0. If we go backwards to equation (1.48) one can see that the photon momentum in vacuum is given by $p=h\nu/c$, or in the vector notation as:

$$\mathbf{p}=\hbar\mathbf{k} \quad , \qquad (1.52)$$

where $\hbar = h/2\pi$ is the reduced Planck constant.

REFERENCES

1. Frank L Pedrotti, Leno M Pedrotti, and Leno S Pedrotti. *Introduction to Optics (3rd Edition)*. Harlow: Pearson Education Limited, 2014.
2. Alex Montwill and Ann Breslin. *Let there be Light*. London: Imperial College Press, 2008.
3. F. Graham Smith, Terry A. King, and Dan Wilkins. *Optics and Photonics: An Introduction (2nd Edition)*. Chichester: John Wiley & Sons Ltd., 2007.
4. Casper Hakfoort. *Optics in the age of Euler*. Cambridge: Cambridge University Press, 1995.
5. Eugene Hecht. *Optics (5th Edition)*. Harlow: Pearson Education Limited, 2017.

6. David Greene. *Light and Dark*. Bristol and Philadelphia: Institute of Physics Publishing Ltd., 2003.
7. Steven Beeson and James W. Mayer. *Patterns of Light*. New York: Steven Beeson and James W. Mayer, 2008.
8. James M Overduin and Paul S Wesson. *The Light/Dark Universe: Galactic Light, Dark Matter and Dark Energy*. Danvers: World Scientific Publishing Co. Pte. Ltd., 2008.
9. Ian R. Kenyon. *The Light Fantastic: A Modern Introduction to Classical and Quantum Optics*. Oxford: Oxford University Press, 2008.
10. Chandrasekhar Roychoudhuri, A. F. Kracklauer, and Katherine Creath (eds). The *Nature of Light: What Is a Photon?* Boca Raton: Taylor & Francis Group, LLC, 2008.
11. Hartmann Römer. *Theoretical Optics: An Introduction*. Weinheim: WILEY-VCH Verlag GmbH & Co. KGaA., 2005.
12. Richard J Weiss. *A Brief History of Light and Those That Lit the Way*. Singapore: World Scientific, 1995.

2
CHAPTER

OPTICAL RADIOMETRY AND PHOTOMETRY

INTRODUCTION

Radiometry is the substantial part in the field of optical measurement and engineering, where the measurement techniques for power, polarization, spectral content and other relevant parameters for an electromagnetic radiation interrogation, which are important in characterization of optical sources and detectors, have been studied. An instrument that measures optical radiation is named a radiometer. Radiometer is typically used to detect the radiation and in many applications, where it is essential, to measure the amount of the captured radiation. As a matter of fact, the word "radiometry" itself means the radiation measurement. The term radiometer is usually exclusively applied to the devices which measure radiance. However, this term will be used in a more general sense for a device that measures one of several power–dependent optical quantities. For example, radiance, defined as the optical power emitted from an area within particular solid angle, is one of many optical terms that will be defined and treated in this chapter. Moreover, this chapter will include both the corresponding mathematical techniques for calculating the radiative transfer as well as the measurement techniques for characterizing of fluxes and radiometric properties of different kinds.

Before starting introducing radiometric quantities there are some important geometrical parameters that must be firstly introduced. One of these parameters is surely the solid angle. A solid angle can be understood as the two–dimensional equivalent of a linear angle or as the two–dimensional angle in three–dimensional space. This angle is the measure of the size of the object that appears to an observer looking at it from a certain point. The solid angle can be equivalently defined as the projection of an area onto a sphere surface divided by the square of the radius of the sphere, as it is presented in figure 2.1:

$$\Omega = \frac{S}{r^2}, \qquad (2.1)$$

where S is the projected area onto the sphere surface and r is the radius of the sphere. The maximal value of the solid angle is equal to 4π sr (steradians).

The way that two bodies exchange the radiated powers or energies to and from each other is described by their mutual radiative transfer. There are several physical quantities that can give more insight in this process, such as, for example, the energy, the power, and certain geometric characteristics of the power, like its density, i.e. the power emitted by the unit area. In order to understand these transfer processes below it will be given some important definitions. The radiometric and photometric quantities, which are listed in table 2.1. together with their typical symbols and units, can be introduced as presented in figure 2.2.

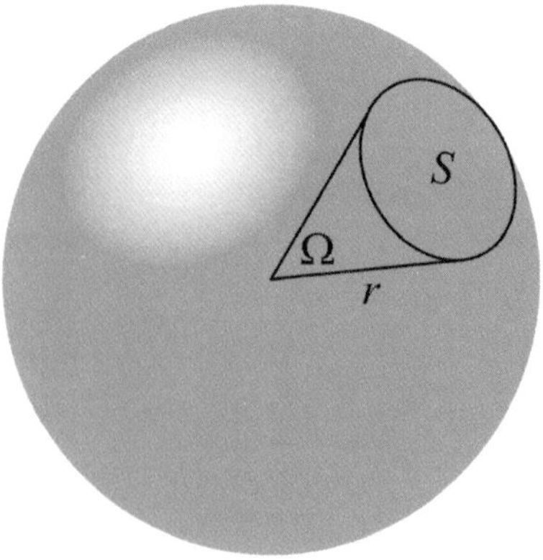

Figure 2.1. Relevant geometrical parameters required for the solid angle definition.

Table 2.1. Radiometric and photometric quantities and their symbols and units

Radiometric quantity	Symbol	Units	Photometric quantity	Symbol	Units
Radiant energy	Q	J	Luminous energy	Q_V	lm s
Radiant flux (power)	Φ, P	W	Luminous flux	Φ_V	lm
Irradiance	E	W/m^2	Illuminance	E_V	$(lm/m^2) = lx$
Radiance	L	$W/(m^2\ sr)$	Luminance	L_V	$lm/(m^2\ sr)$
Radiant intensity	I	W/sr	Luminous intensity	I_V	(lm/sr) = cd
Radiant exitance	M	W/m^2	Luminous exitance	M_V	lm/m^2
Radiant exposure	H	$W\ s/m^2$	Luminous exposure	H_V	Lx s
Radiance temperature	T	K	Color temperature	T_C	K

Optical power or radiant flux Φ, emitted by the light source or detected by the detector, is the primary radiometric quantity measured by radiometers. In some cases flux is the radiometric parameter to be measured, but in many other radiometric measurements radiance of the source or irradiance incident upon a surface are the quantities of interest. According to figure 2.2., an elementary area dA_S acts as a source thus emitting the radiation toward the elementary area dA_R acting as the receiver. In order to simplify the description, it will be assumed that the area dA_S emits uniformly in all directions and that both surfaces are mutually parallel and cantered on and perpendicular to the centreline. For a source of radiation that emits uniformly in all directions it is said that it acts as a Lambertian source. Following this approach, the radiant intensity can be presented as the radiant flux from a centre point of the elementary area dA_S into the cone of light with the base located onto the elementary area dA_R. Therefore, the radiant intensity can be defined as the elementary radiant flux emitted into the elementary solid angle. This radiometric quantity is typically used with the point sources, but in general, it designates the radiant flux per unit solid angle from an entire area:

$$I = \frac{d\Phi}{d\Omega}, \qquad (2.2)$$

where I represents the radiant intensity and $d\Phi$ represents the elementary radiant flux emitted in the elementary solid angle $d\Omega$. The radiance of the source is defined as the radiant flux from the elementary area dA_S of the source within the space defined by the truncated cone of radiation emitted from the entire elementary area dA_S that impinge the receiver elementary area dA_R. Therefore, the radiance L can be defined as the elementary radiant flux emitted into the elementary solid angle from the elementary area as:

$$L = \frac{d\Phi}{dA d\Omega}. \qquad (2.3)$$

where dA is the elementary area. If it is assumed that the radiant flux, which impinges the receiver elementary surface dA_R, is uniformly distributed over the entire area, the irradiance E is defined as the elementary radiant flux per elementary area, i.e. as:

$$E = \frac{d\Phi}{dA}. \qquad (2.4)$$

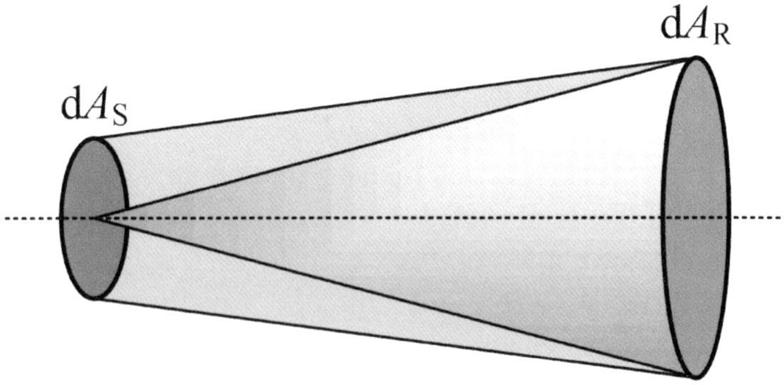

Figure 2.2. Geometric representation of basic radiometric quantities.

OPTICAL RADIOMETRY

Radiative transfer between the bodies is of the paramount importance in the optical radiometry as it explains the concepts of energy and power transfer between the surfaces as well as in the optical systems. In order to determine the transfer of radiation between the arbitrary surfaces the radiometric quantities such as radiant flux, radiant intensity, radiance, and irradiance will be taken into account.

The Radiative Transfer

In order to determine the amount of the radiant flux that originates from the elementary area on one body that reaches the elementary area of the other body we will observe the position of two corresponding radiating objects in the space with all relevant geometrical parameters for the analysis that are presented in figure 2.3. Radiance is the typical parameter that describes the radiating characteristics of a surface. Therefore, it will be assumed that the radiance of the first object at the place of the elementary area dA_1 is equal to L_1. Consequently, the radiance of the second object at the place of the elementary area dA_2 is equal to L_2. The distance between the elementary surfaces is R. The angles between the unit vectors $\mathbf{n}_1$ and $\mathbf{n}_2$, normal to the elementary areas dA_1 and dA_2, respectively, and the line that connects elementary areas centres are θ_1 and θ_2, respectively.

The elementary fluxes $d\Phi_1$ and $d\Phi_2$, radiated by the elementary areas dA_1 and dA_2, respectively, in the direction of the elementary areas dA_2 and dA_{21}, respectively, are given by:

$$\begin{aligned} d\Phi_1 &= L_1 \cos\theta_1 dA_1 d\Omega_2 \\ d\Phi_2 &= L_2 \cos\theta_2 dA_2 d\Omega_1 \end{aligned} \quad (2.5)$$

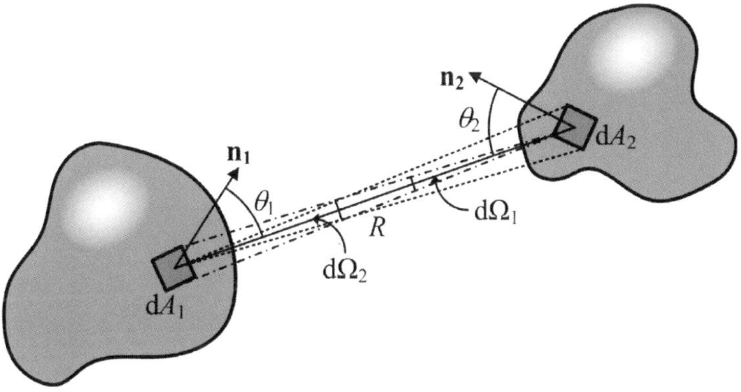

Figure 2.3. Position of two radiating objects in the space with all relevant geometrical parameters for the analysis.

Based on the geometry presented in figure 2.3., for the solid angles $d\Omega_1$ and $d\Omega_2$, i.e. the solid angles under which one elementary has been seen from another, the following is valid:

$$d\Omega_1 = \frac{\cos\theta_1}{R^2} dA_1$$

$$d\Omega_2 = \frac{\cos\theta_2}{R^2} dA_2 \quad . \qquad (2.6)$$

By combining equations (2.5) and (2.6) we have:

$$d\Phi_1 = \frac{L_1 \cos\theta_1 \cos\theta_2}{R^2} dA_1 dA_2$$

$$d\Phi_2 = \frac{L_2 \cos\theta_1 \cos\theta_2}{R^2} dA_1 dA_2 \quad . \qquad (2.7)$$

The radiant flux $d\Phi_1$ ($d\Phi_2$), given by equation (2.7), represents also the radiant flux originating from the elementary area dA_1 (dA_2) that is captured by the elementary area dA_2 (dA_1).

The Lambertian Emitters

The Lambertian radiator has the angle independent radiance, i.e. its radiance does not depend on the angle into which the radiation is directed, so $L \neq L(\theta)$ is valid. As the Lambert emitter is the isotropic radiator the determination of the radiant exitance will be presented on the base of the geometrical parameters given in figure 2.4. An elementary area dS, positioned onto the radiating body that obeys Lambert law, radiates with the constant radiance L into the overlying hemisphere with the radius R. According to equation (2.7), the elementary radiant flux $d^2\Phi$, emitted into the cone limited by the azimuth angles $[\varphi, \varphi+d\varphi]$ and by the elevation angles $[\theta, \theta+d\theta]$, is given by:

$$d^2\Phi = \frac{L\cos\theta\cos\psi}{R^2} dS dA \quad , \qquad (2.8)$$

where $\psi = 0$ as the normal to the elementary area dA, positioned onto the hemisphere surface, is directed toward its centre, i.e. toward the elementary area dS. Based on the geometrical parameters in figure 2.4., the elementary area dA is equal to $dA = R^2 \sin\theta d\theta d\varphi$. Therefore, for the elementary radiant flux $d^2\Phi$ the following is valid:

$$d^2\Phi = L\sin\theta\cos\theta dS d\theta d\varphi . \quad (2.9)$$

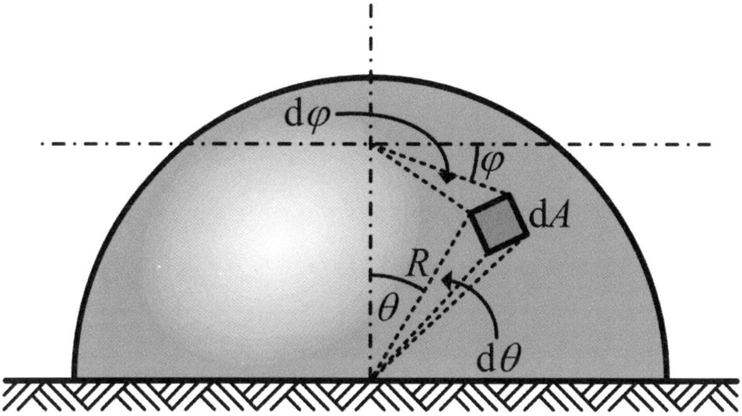

Figure 2.4. Lambertian emitter relevant geometry.

The radiant exitance M, or equivalently the radiant flux per unit area is thus given as:

$$M = \frac{d\Phi}{dS} = L \int_{\theta=0}^{\theta=\pi/2} \int_{\varphi=0}^{\varphi=2\pi} \sin\theta\cos\theta d\theta d\varphi . \quad (2.10)$$

Integration of equation (2.10) gives:

$$M = \pi L . \quad (2.11)$$

The radiant exitance is π times the radiance, or as usually said, the radiance is the radiant exitance divided by π.

Radiometric Measurements

All radiometric quantities that have been introduced, such as energy, radiant flux, radiance, etc., are integral radiometric quantities. Measurement of such energetic quantities can be performed by using some kind of thermal detectors that converts the captured energy into the heat and subsequently into the temperature increase of the detector, such as thermocouples and bolometers. However, the sensitivity of such detectors is, rather low so the measurements of such parameters are usually performed by the optical detectors which have different sensitivities for different wavelengths such as solid state photodetectors (photodiodes, avalanche photodiodes, photoconductors) and photomultipliers. Typically, silicon photodiode are sensitive in the range from 200 nm to 1100 nm with the peak sensitivity in the near infrared range (800 – 900 nm). As the radiometric measurement can be spectrally sensitive one can define the corresponding spectral radiometric quantities. The spectral radiant flux Φ_λ or Φ_ν is the radiant flux emitted by the source in the wavelength range [λ, λ+dλ] or in the frequency range [ν, ν+dν] given by:

$$\Phi_\lambda = \frac{\mathrm{d}\Phi}{\mathrm{d}\ddot{e}}$$

$$\Phi_\nu = \frac{\mathrm{d}\Phi}{\mathrm{d}\nu} \quad (2.12)$$

Bearing in mind that $\lambda\nu = c$ we have:

$$\Phi_\lambda = \frac{\mathrm{d}\Phi}{\mathrm{d}\ddot{e}} = \frac{\mathrm{d}\Phi}{\mathrm{d}\left(\frac{c}{\nu}\right)} = -\frac{\nu^2}{c}\frac{\mathrm{d}\Phi}{\mathrm{d}\nu} = -\frac{\nu^2}{c}\Phi_\nu \quad (2.13)$$

Typical configuration of the radiometric measurement is given in figure 2.5., where the photodetector with spectrally dependent sensitivity is used to characterize the source of radiation. A source of optical radiation emits the radiation through the source aperture of area A_S. The detector aperture of the area A_D is positioned at the distance r from the source aperture. The area of the detector aperture limits the flux that after passing through the filter F impinge the detector D. In the case of a solid–state detector the output signal is an electrical current i_D which is proportional to the detector incident flux Φ_D with the conversion factor $\Re$, i.e. $i_D = \Re\Phi_D$. In order to convert the current signal into the voltage signal a transimpedance amplifier A is used. The relation that describes the overall signal response of the measurement chain is given by:

$$v(\lambda) = R(\lambda)\Phi(\lambda) \quad (2.14)$$

where $v(\lambda)$ is the output voltage signal for a given wavelength λ and $R(\lambda)$ is the overall responsivity of the measurement system.

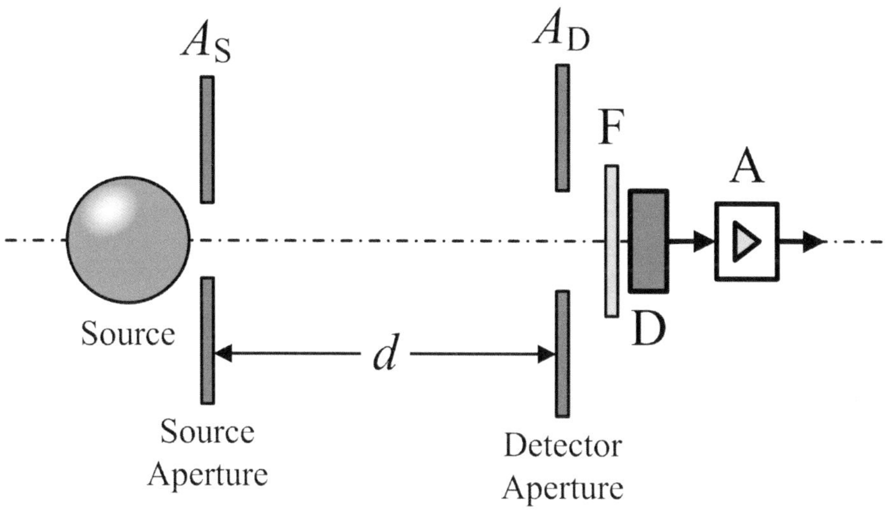

Figure 2.5. Typical configuration of the radiometric measurement system.

The relation between the output signal of the entire measurement system, given in figure 2.5., and the relevant parameters of the same will be derived in the case of the radiation source that follows the Lambertian law with the spectral radiance of $L_\lambda(\lambda)$. Typically, the apertures are of the circular shape, normal to the line of sight, with the radiuses of r_S and r_D for source and detector apertures, respectively. The distance between the apertures is d. In the case of uniform spatial distribution of the detector sensitivity the output voltage signal v is given by:

$$v = \frac{A_S A_D (1+c)}{r_S^2 + r_D^2 + d^2} \int_\lambda L_\lambda(\lambda)\tau(\lambda)R(\lambda)\mathrm{d}\lambda \quad , \qquad (2.15)$$

where c is the correction factor that depends on the system geometry and $\tau(\lambda)$ is the filter transmittance. Typical employed radiant sources have the spectrum located at the wavelength λ_0 with the spectral width $\Delta\lambda$ that is much smaller than the central frequency $\Delta\lambda << \lambda_0$, or the band pass of the filter is narrow compared to the central frequency. In this case equation (2.15) can be simplified into:

$$v \approx \frac{A_S A_D (1+c)}{r_S^2 + r_D^2 + d^2} L_\lambda(\lambda_0)\tau(\lambda_0)R(\lambda_0)\Delta\lambda \quad . \qquad (2.16)$$

The last equation is often used in radiometry but it is important to be aware of the assumptions that have been made in order to derive it.

Wavelength selective measurement of the optical radiation is performed by a spectral radiometer. The measurement system, presented in figure 2.5., although aimed for radiometric measurement at the particular wavelength, defined by the optical filter, cannot be referred as the spectral radiometer as the measurement is performed at the fixed wavelength. The measurement principle of the spectral radiometer is given in figure 2.6. aimed for spectral radiance measurement. The measurement system is presented in such a way that with minimal changes in the design it can be adapted to measure spectral irradiance or spectral flux. Similarly to the measurement system in figure 2.5., the wavelength selective measurement system measure the radiation from the source that passes through the source aperture of area A_S, detector aperture of area A_D. After passing the lens L the optical radiation is focused onto the monochromator M that at its entrance has a slit S_1 and an exit slit S_2. A monochromator is an optical device that can transmit a selectable narrow band of wavelengths of optical radiation from a wider range of wavelengths that illuminates its entrance. Due to the dispersive element of the monochromator the optical radiation with the selected wavelength impinge the detector D that is further connected to the signal amplifier A. The overall spectral responsivity $R_\lambda(\lambda_0)$ of the imaging system, the monochromator, and the detector is a function of wavelength where the monochromator is adjusted to the wavelength λ_0. Dependence of transmittance, reflectance of the used optical components can also influence the overall spectral responsivity.

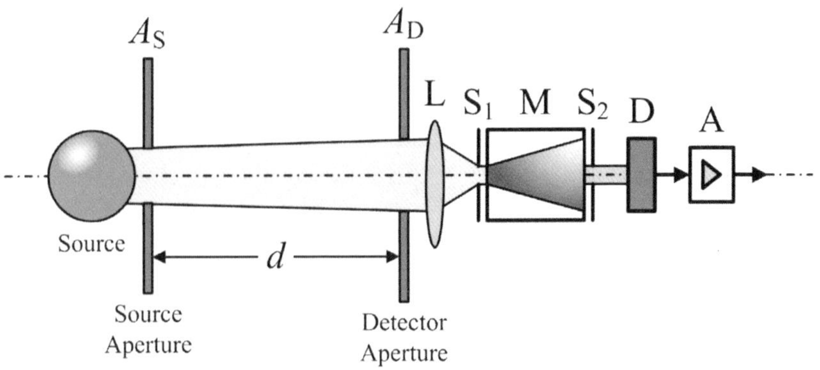

Figure 2.6. Measurement principle of the spectral radiometer.

Internationally recognized designations for the various wavelength regions and its wavelength intervals that are commonly used in radiometric measurements are presented in table 2.2.

Table 2.2. Wavelength regions and wavelength intervals.

Wavelength regions	Wavelength intervals
UV–C	100 – 280 nm
UV–B	280 – 315 nm
UV–A	315 – 400 nm
Visible	380 – 780 nm
IR–A	780 – 1400 nm
IR–B	1.4 – 3 μm
IR–C	3 – 1000 μm

Reflection, Absorption, and Transmission

As mentioned in the previous chapter transmission, reflection, and absorption, which can be caused by various factors, can significantly influence the overall responsivity and spectral responsivity of the radiometric measurement system. Generally, a material may transmit, reflect, emit or absorb radiation, and typically does more than one of these at a time. Spectral and geometric parameters of an optical component influence all of these properties. The corresponding processes are graphically presented in figure 2.7. for a light incident on a plane parallel plate were it is shown that the incident light can be reflected, absorbed or transmitted through the material.

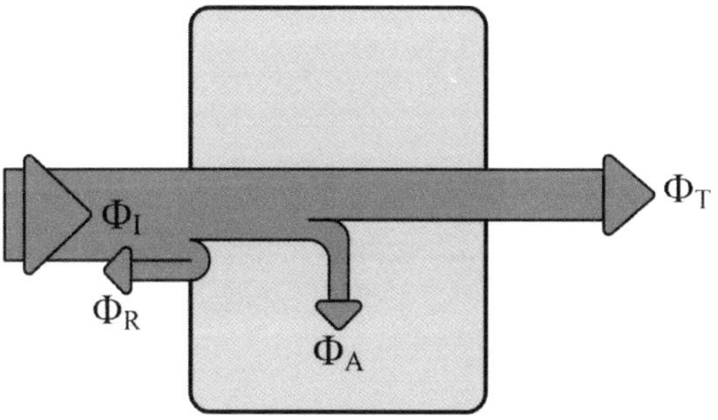

Figure 2.7. Transmittance, absorptance, and reflectance.

Based on the energy conservation law, the sum of reflected Φ_R, absorbed Φ_R, and transmitted Φ_T flux must be equal to the incident flux Φ_I or:

$$\Phi_R + \Phi_A + \Phi_T = \Phi_I$$

$$\frac{\Phi_R}{\Phi_I} + \frac{\Phi_A}{\Phi_I} + \frac{\Phi_T}{\Phi_I} = \rho + \alpha + \tau = 1, \qquad (2.17)$$

where the reflectivity ρ, absorptivity α, and transmissivity τ are defined by this equation. If there are no nonlinear processes within the material and thus no change in the radiation frequency (there is no frequency doubling, no fluorescence, etc.) the energy conservation law can be also applied to any particular wavelength, so we have:

$$\frac{\Phi_R(\lambda)}{\Phi_I(\lambda)} + \frac{\Phi_A(\lambda)}{\Phi_I(\lambda)} + \frac{\Phi_T(\lambda)}{\Phi_I(\lambda)} = \rho(\lambda) + \alpha(\lambda) + \tau(\lambda) = 1, \qquad (2.18)$$

A similar argument can be applied to each of the linear or circular polarization components, as long as the polarization is conserved.

Kirchhoff's Law

Kirchhoff's law of heat radiation states that under certain conditions the emissivity is equal to the absorptivity for isothermal bodies. If we consider an isothermal cavity that is also in thermal equilibrium, as shown in figure 2.8., which every part is at the same constant temperature T and the walls are made of the same material, meaning that optical properties of the inner cavity walls are uniform, then according to the energy conservation law the cavity walls radiant flux of the emitted radiation must be equal to the flux that is captured by the cavity walls. Equilibrium is maintained only if the radiation is perfectly reflected back. If the radiation doesn't perfectly reflect then any deficiency in the reflectivity must be compensated exactly by emissivity, or:

$$\varepsilon = 1 - \rho, \qquad (2.19)$$

where ε is the emissivity. Then, by combining equations (2.17) and (2.19) and taking into considerations that the cavity walls are opaque ($\tau = 0$), we have:

$$\varepsilon = \alpha, \qquad (2.20)$$

or the emissivity must be equal to absorptivity. Similarly, it can be stated that for each wavelength the emissivity is equal to absorptivity:

$$\varepsilon(\lambda) = \alpha(\lambda), \tag{2.21}$$

while in the opposite case ($\varepsilon(\lambda) \neq \alpha(\lambda)$) the energy conservation law would be broken.

Figure 2.8. Isothermal cavity.

However, there are some cases where the application of the Kirchhoff's law is not straight forward. For example when bodies are covered with the paints with metallic particles, layered optical materials, or semi-infinite bodies which has a large thermal gradient at the surface.

MEASUREMENT TECHNIQUES IN RADIOMETRY

Absolute Radiometer

Absolute radiometer is an instrument that can, not only measure the level of captured optical radiation, but cannot be simply calibrated by comparing it with another instrument or reference source, i.e. it must be a self–calibrated device. The absolute radiometer is typically based on one of two basic physical principles: electrical substitution (thermal effects occur) and quantum physics (photon detection). In order to make a solid state detector an absolute radiometer one can perform several measurements to find the quantum efficiency of the solid state detector (photodiode) i.e. the number of charge carriers that arrive at the inputs of the measuring electrical circuitry per incident photon, and thus to estimate its spectral response. The photodiode responsivity $\Re_\lambda$ at a particular wavelength λ is given as:

$$\Re_\lambda = \frac{e}{h}\eta \quad , \qquad (2.22)$$

where e is the electron charge and η is the photodiode quantum efficiency. The basic principle that lies behind the photodiode as the absolute radiometer is the fact that quantum efficiency of the photodiode can be determined from a number of simple experiments. In order to accurately determine the photodiode efficiency one must measure its reflectance ρ as it is satisfied $\eta = 1\text{-}\rho$. The photodiode reflectance can be easily measured very accurately.

Beside the radiometers based on the quantum effects, as mentioned above there are also absolute radiometers based on the electrical substitution principles. The operating principle of electrical substitution radiometers is to compare the heating effect of optical radiation with the heating effect of the corresponding electrical heater. Figure 2.9. presents the operating principle of the electrical substitution radiometer. A typical electrical substitution radiometer consists of an optical absorbing element, which simplest construction can be in the shape of a metal plate, connected by the poorly conducting heat link, to a reference heat sink, which is kept at a constant temperature. The maintenance of the constant temperature of the heat sink can be accomplished actively by a temperature control loop, or passively through its thermal mass.

The main physical principle is to keep the same temperature at the reference heat sink whether it is exposed to the optical radiation with the radiant flux Φ_O or heated by means of an electrical heater that delivers power P_E to the reference heat sink. The temperature difference ΔT between the ambient and the reference heat sink is given by:

$$\Delta T_O = K(\Phi_O + \Phi_B)$$
$$\Delta T_E = K(P_E + \Phi_B) \quad , \qquad (2.23)$$

where ΔT_O is the temperature difference caused by the optical radiation, ΔT_E is the temperature difference caused by the electrical heater, K is the constant that depends on the radiometer construction, and Φ_B is the radiant flux of the background radiation. By keeping $\Delta T_E = \Delta T_O$ one can simply measure the radiant flux by measuring the delivered electrical power as $\Phi_O = P_E$.

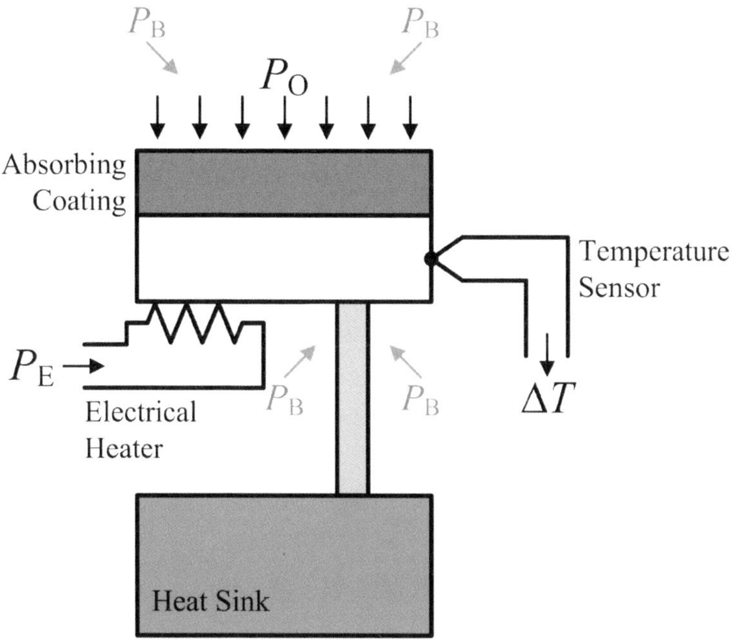

Figure 2.9. Block schematic representation of the absolute radiometer based on the electrical substitution principle.

Radiant Flux Measurement

The radiometric quantities measurements, such as radiant flux, irradiance, radiance, and intensity are made with the help of a radiometer, i.e. the instrument that measures the optical flux captured by its surface. Typically, the flux that impinges the detector surface generates an electrical signal, which is proportional to the captured flux. By calibration it is possible to make a relation between the measured radiant flux and the corresponding electrical signal. It is important to notice that due to the relation between the radiant flux and certain geometrical parameters it is possible to measure other radiometric quantities by involving appropriate adjustments of the measurement setup. For example, if the detector measures a certain radiant flux that is captured by the detector active surface, the irradiance or the flux density can be measured by dividing the measured flux by the active area of the detector surface. Of course, the measured irradiance is averaged over the area of the entrance aperture.

Exitance is similar to the irradiance but it the property of the radiation source. However, it can be also determined by measuring the radiant flux that is captured by the detector. If the radiant flux is known then, based on the geometry of the measurement setup, the exitance of the source can be determined if the distance from the source to the detector and the size of the source are known.

In order to measure the radiant intensity of a source the radiant flux angular distribution must be determined. Typically, in the case of a point source radiant intensity can be measured by measuring the optical flux that impinges on the detector and certain geometrical parameters of the measurement setup. First of all, the distance between the source and the detector must be much larger than the source dimensions thus "forcing" the source to be a point like source. If the distance is large enough, as it is presented in figure 2.10., the radiation emitted in the solid angle Ω defined by the detector aperture area A and the distance d as $\Omega = A/d^2$ will be captured by the detector D and amplified by the signal amplifier A, so the measured flux will be Φ. According to the presented geometry of the measurement setup, the measured radiant intensity will be:

$$I = \frac{\Phi}{\Omega} = \frac{d^2}{A}\Phi, \qquad (2.24)$$

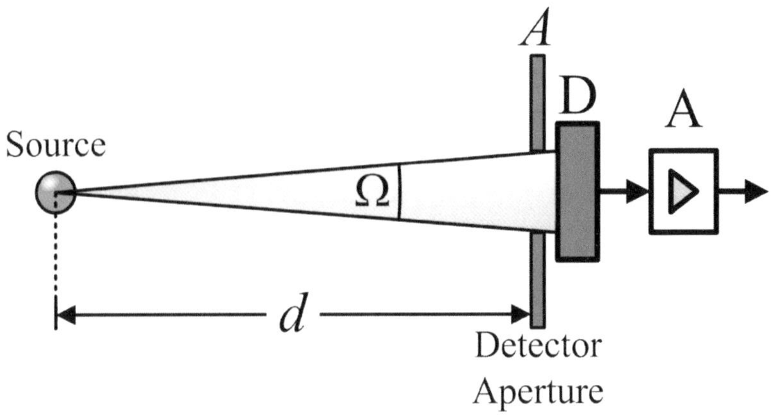

Figure 2.10. Measurement setup for measuring the radiant intensity.

The radiance measurement can be performed according to the measurement setup presented in figure 2.11. The radiation source and the detector D are positioned in such a way the between them, in the simplest measurement arrangement, is an aperture with the entrance opening area A_1. This aperture serves to limit the solid angle to a certain value to enable that all of the radiations that pass through the aperture impinges the detector. This condition is fulfilled if it is satisfied:

$$\frac{A_1}{d_1^2} < \frac{A_2}{(d_1 + d_2)^2}, \tag{2.25}$$

where d_1 is the distance between the aperture and the source and d_2 is the distance between the aperture and the detector and A_2 is the entrance opening area of the detector aperture.

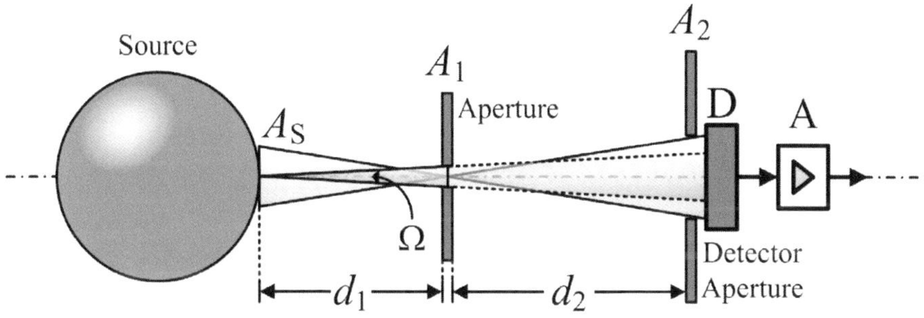

Figure 2.11. Measurement setup for measuring the radiance.

In order to avoid the averaging processes $d_1^2 < A_1$ and $(d_1 + d_2)^2 < A_2$ must be fulfilled, i.e. all solid angles must be much smaller than unity. If the detector D and signal amplifier A measure the radiant flux of Φ then the corresponding radiant intensity of the source at a given surface point is given by:

$$I = \frac{\Phi d_1^2}{A_1}. \tag{2.26}$$

Due to very small opening of the aperture between the source and the detector and the corresponding geometrical parameters of the measurement setup, the detector receives the radiation only from a small portion of the source surface. The area of the source that emits the light toward the detector is given by:

$$A_S = \frac{d_1}{d_2} A_2 \quad . \tag{2.27}$$

According to the equations (2.26) and (2.27) the measured radiance is given by:

$$L = \frac{I}{A_S} = \frac{d_1 d_2}{A_1 A_2} \Phi \quad . \tag{2.28}$$

The presented measurement schematic, aimed for the radiance measurements, is given in its simplest form, i.e. where are no need for additional optical elements. For example, if instead of a simple aperture between the source and the detector a lens with much higher aperture area is placed, then a much higher levels of captured optical power can lead to the higher precision measurement. One must pay attention that all solid angles still must be kept well below unity in order to avoid averaging. Typical measurement setup with the lens in the middle is given in figure 2.12.

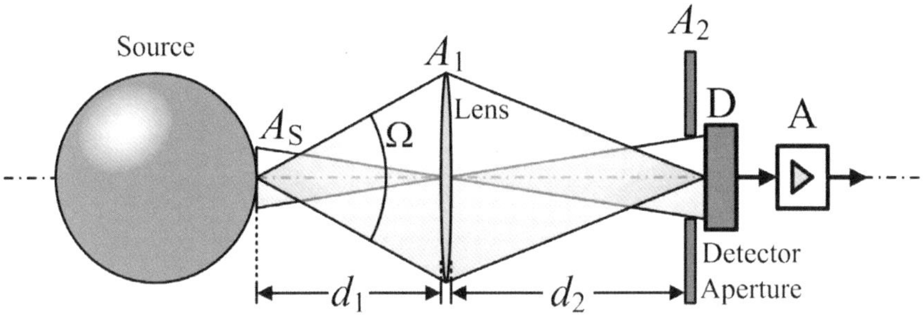

Figure 2.12. Measurement setup for measuring the radiance with incorporated simple lens based optical system.

The measurement of low radiant fluxes emanating from faint radiation sources can be the problem if the measurement has been performed with the sensing setups shown above. Due to the very low light levels the influences of other parasitic effects can interfere and thus hide the useful signal. For example, relative high levels of background light, voltage offset of the signal amplifier and its thermal and time drifts as well as high levels of flicker noise of the amplifier can significantly deteriorate the measurement accuracy and precision. However, there is a solution in moving the useful signal toward the higher frequencies and perform the phase sensitive detection with the help of lock–in amplifier. The typical measurement setup is presented in figure 2.13. Between the source and the main detector D_1 an optical chopper has been installed. An optical chopper is a device that periodically interrupts a light beam. There are several different types of optical choppers. One of them is presented in figure 2.13. and consists of an opaque rotating wheel with symmetrically positioned transparent parts (holes) that is driven by a motor M. The signal from the detector is amplified by the signal amplifier A_1 which output has been brought to the input port of the lock–in amplifier LA. In order to perform a phase sensitive detection there is a need for the corresponding reference signal which must be brought to the reference input of the lock–in amplifier. The reference signal is obtained with the help of auxiliary light source LS and detector D_2 that are separated by the same optical chopper wheel. The reference signal is obtained at the output of the second signal amplifier A_2. The lock–in amplifier output signal will be proportional to the detector captured radiant flux but most of the disruptive influences will be significantly suppressed.

Accurate and high precision measurement of very low light levels or radiant fluxes draws a lot of attention in the past two decades as one of the most powerful measurement techniques in order to detect and find the extrasolar planets. The main principle of an application of the above presented measurement technique is given in figure 2.14. When the planet transits the star disk in the direction A–B–C–D–E there is a small drop in the measured radiant flux emanating from the star, as presented on a diagram given in the lower part of figure 2.14. The reason for this phenomenon is the shadow that the planet makes. The measured flux when the planet is far away from the star Φ_0 is just slightly higher than in the case when the planet transits the star disk Φ_0-$\Delta\Phi$, where $\Delta\Phi$ is the flux drop that is typically much smaller than the star flux $\Delta\Phi<<\Phi_0$. As the transition last for a very long period of time, the measurement technique that is capable of detecting extrasolar planets must be very precise in order to be able to measure very small flux changes and very stable (low output signal drift) for a very long period of time (suppressed flicker noise).

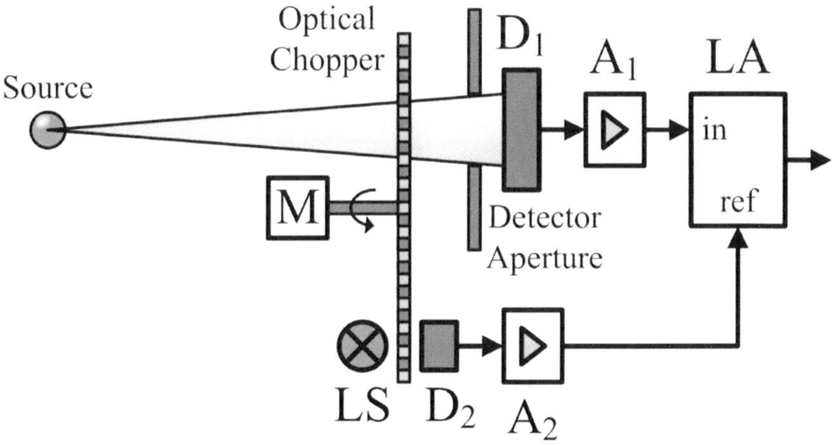

Figure 2.13. Measurement setup aimed for measuring low radiant fluxes.

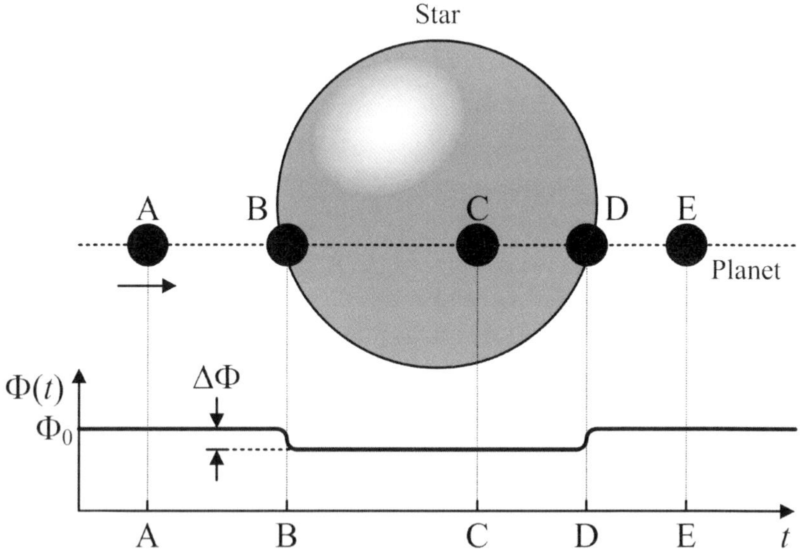

Figure 2.14. Method for detecting the extrasolar planet.

Integrating Sphere

An integrating sphere is an optical device that consists of a spherical shell which inner walls are covered with a diffuse, white and highly reflective coating. In order to be able to perform the measurements there are small holes at the sphere for entrance and exit ports. The main function of an integrating sphere is to spatially integrate the radiant flux by a uniform scattering or diffusing effect that occurs at its inner walls. Light rays, which are incident on any point on the integrating sphere inner walls, are due to the multiple scatterings, distributed evenly to the all other points at the inner sphere surface. An integrating sphere is typically used with some light source and a detector for radiant flux measurement and it has become a standard instrument in optical radiometry and photometry. It has a unique function of the total flux measurement of any light source in a single measurement. Typical integrating sphere arrangement is presented in figure 2.15. The light, which radiant flux Φ_I is to be measured, enters the integrating sphere through the input hole with the area of entrance opening of A_I. Due to the multiple scatterings at the inner highly reflective and diffusive walls the light emanating from the source illuminates the inner walls with uniform irradiance E.

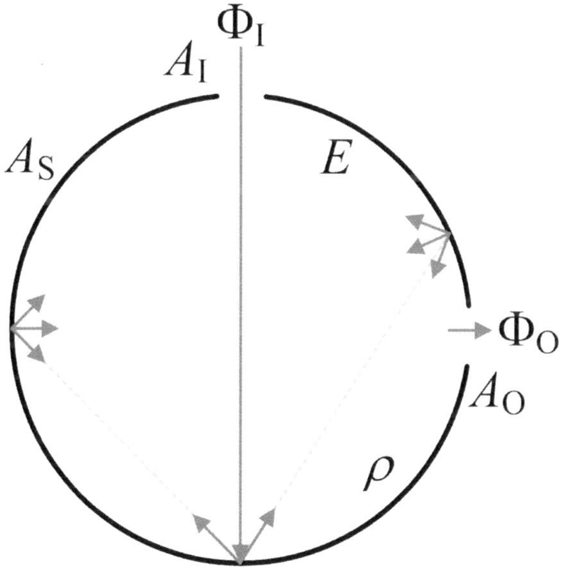

Figure 2.15. Integrating cavity with relevant parameters.

Based on the energy conservation law we have:

$$\Phi_I = E(A_I + A_O) + E(1-\rho)(A_S - A_I - A_O), \qquad (2.29)$$

where A_O represents the area of the exit opening, ρ is the reflectance of the inner walls, and A_S is the overall area of the inner walls. One can see that the incoming flux is divides into two parts. The first one, which is represented by the first argument at the right side of equation (2.29), is the flux loss due to the entrance and exit holes where the light can leave the integrating sphere. The second one, which is represented by the second argument at the right side of equation (2.29), is the flux loss due to the wall absorption. If the detector is positioned at the exit hole and captures all the flux leaving the sphere at that point, the detector measures the flux Φ_O which is given by:

$$\Phi_O = EA_O. \tag{2.30}$$

By combining the last two equations, we have:

$$\Phi_O = \frac{\Phi_I}{\left(1+\frac{A_I}{A_O}\right)+(1-\rho)\left[\frac{A_S}{A_O}-\left(1+\frac{A_I}{A_O}\right)\right]}, \tag{2.31}$$

i.e. the measured flux is proportional to the incoming flux (light source flux). Typically, the integrating sphere is calibrated by the known radiant flux Φ_{IC}, where the measured flux is Φ_{OC}, so, based on the measured flux Φ_O one can determine the light source flux as:

$$\Phi_I = \frac{\Phi_C}{\Phi_O}\Phi_O = K_C\Phi_O, \tag{2.32}$$

where $K_C = \Phi_C / \Phi_O$ is the calibration constant. One can notice that integrating sphere parameters such the areas of holes, the area of integrating sphere inner walls, and the reflectance of the walls, although figure in the measurement, are not necessary to know as the calibration process eliminates them from the equation. As mentioned previously, an integrating sphere is capable of measuring the total radiant flux of an arbitrary light source. Simply, by placing the source inside the integrating sphere and hiding it behind the baffles that are covered with the same material as the inner sphere walls, the measured output flux will be proportional to the total source flux. The baffles serve to stop the source radiation to left the integrating sphere before its reflection from the walls.

PHOTOMETRY

In photometry the light is measured in such a way that the measurement result is in accordance with the human vision. A large number of devices are aimed for human use such as computer displays, TVs, traffic signal lights, etc., therefore they must be tested and evaluated based on the spectral responsivity of the average human eye. Radiometry covers all optical spectral regions, ranging from ultraviolet to infrared, while the photometry only the range of light that is visible for average human eye, i.e. in the spectral range from 380 nm to 770 nm. In the past times human eye was used exclusively as an optical detector. In order to measure the intensity of a test light, the test light source was moved and placed at the varied distances from an eye or a screen and compared with the intensity of a standard light source positioned at a fixed position by a visual observation. The distance of the test light source was set so that both light sources, test and standard, appear equally bright. The estimated intensity of the test light source is calculated as the ratio of the intensity of the standard source and the ratio of the distances squared. Such a measured light intensity was named candle power, or in present terminology luminous intensity. This photometric quantity was the first defined photometric quantity. Since 1940s the human eye, which was predominantly used as an optical detector, has been replaced by a light sensitive detectors (solid–state detector, photomultiplier tube).

Modern photometry measurements have been performed with the help of photodetectors, thus referring to the physical photometry. Physical photometry characterizes the optical radiation with an optical detector that mimics the spectral response of the human eye, i.e. it has incorporated weighted spectral response same as the spectral response of a human eye. According to the table 2.1., the photometric units include the lumen or the luminous flux, the candela or the luminous intensity, the lux or the illuminance, and the candela per square meter or the luminance. Each photometric quantity is the spectrally integrated corresponding radiometric quantities weighted by the human eye spectral response.

Spectral Response of a Human Eye

In 1924 the relative spectral response of the human eye was specified by the Commission Internationale de l'Eclairage (CIE), (the International

Commission on Illumination). Such a defined relative spectral response of the human eye was called the spectral luminous efficiency for photopic vision with a joined symbol $V(\lambda)$. The relative spectral response is defined in the spectral domain ranging from 360 nm to 830 nm and it is normalized to its peak value at 555 nm. The values were republished by CIE in 1983, and adopted by Comité International des Poids et Mesures (CIPM) (the International Committee on Weights and Measures) in the same year. Tabular presentation of the spectral luminous efficiency for photopic vision values at 5 nm increments are presented in table 2.3. One can notice from the tabular values that outside the spectral range ranging from 380 nm to 770 nm the values of the spectral luminous efficiency for photopic vision are smaller than 10^{-4} thus negligible for the calculations

A photometric quantity X_V can be related to the corresponding radiometric spectral quantity X_λ in the following way:

$$X_V = K_m \int_{\lambda=360\mathrm{m}}^{\lambda=830\mathrm{m}} X_\lambda V(\lambda) d\lambda \quad , \qquad (2.33)$$

where K_m is the constant which relates photometric and radiometric quantities and it is named the maximum spectral luminous efficiency of radiation for photopic vision. The value of this constant is defined as the spectral luminous efficiency of radiation at the frequency of 540×10^{-12} Hz (or at the wavelength of 555.016 nm in the standard air) to be K_m = 683 lm/W. This value isn't defined strictly for the wavelength of 555 nm, where the peak value should appear, but for the wavelength that is slightly shifted from the maximum value. The exact maximum value of the constant K_m can be accurately calculated as 683 lm/W $\times V(\lambda = 555$ nm$)/V(\lambda = 555.016$ nm$) = 683.002$ lm/W, which can be simply rounded to 683 lm/W, with the negligible error.

The function $V(\lambda)$, which values are given in table 2.3. and which is also graphically presented in figure 2.16. is defined for the standard photometric observer for photopic vision at relatively high luminance levels (higher than 1 cd/m^2). The spectral responsivity of an average human eye significantly differs at very low luminance levels (at luminance levels lower than 10^{-3} cd/m^2) when the eye rods are the predominant receptors.

This type of vision is named scotopic vision and its spectral luminous efficiency values $V'(\lambda)$ are presented in table 2.4. and graphically in figure 2.16. The scotopic vision spectral luminous efficiency has the maximum value at 507 nm.

Table 2.3. Definitive values of the spectral luminous efficiency function for photopic vision $V(\lambda)$

λ [nm]	V(λ) [a.u.]	λ [nm]	V(λ) [a.u.]	λ [nm]	V(λ) [a.u.]	λ [nm]	V(λ) [a.u.]
360	3.9170×10^{-6}	480	0.13902	600	0.63100	720	1.0470×10^{-3}
365	6.9650×10^{-6}	485	0.16930	605	0.56680	725	7.4000×10^{-4}
370	1.2390×10^{-5}	490	0.20802	610	0.50300	730	5.2000×10^{-4}
375	2.2020×10^{-5}	495	0.25860	615	0.44120	735	3.6110×10^{-4}
380	3.9000×10^{-5}	500	0.32300	620	0.38100	740	2.4920×10^{-4}
385	6.4000×10^{-5}	505	0.40730	625	0.32100	745	1.7190×10^{-4}
390	1.2000×10^{-4}	510	0.50300	630	0.26500	750	1.2000×10^{-4}
395	2.1700×10^{-4}	515	0.60820	635	0.21700	755	8.4800×10^{-5}
400	3.9600×10^{-4}	520	0.71000	640	0.17500	760	6.0000×10^{-5}
405	6.4000×10^{-4}	525	0.79320	645	0.13820	765	4.2400×10^{-5}
410	1.2100×10^{-3}	530	0.86200	650	0.10700	770	3.0000×10^{-5}
415	2.1800×10^{-3}	535	0.91485	655	8.1600×10^{-2}	775	2.1200×10^{-5}
420	4.0000×10^{-3}	540	0.95400	660	6.1000×10^{-2}	780	1.4990×10^{-5}
425	7.3000×10^{-3}	545	0.98030	665	4.4580×10^{-2}	785	1.0600×10^{-5}
430	1.1600×10^{-2}	550	0.99495	670	3.2000×10^{-2}	790	7.4657×10^{-6}
435	1.6840×10^{-2}	555	1.00000	675	2.3200×10^{-2}	795	5.2578×10^{-6}
440	2.3000×10^{-2}	560	0.99500	680	1.7000×10^{-2}	800	3.7029×10^{-6}
445	2.9800×10^{-2}	565	0.97860	685	1.1920×10^{-2}	805	2.6078×10^{-6}
450	3.8000×10^{-2}	570	0.95200	690	8.2100×10^{-3}	810	1.8366×10^{-6}
455	4.8000×10^{-2}	575	0.91540	695	5.7230×10^{-3}	815	1.2934×10^{-6}
460	6.0000×10^{-2}	580	0.87000	700	4.1020×10^{-3}	820	9.1093×10^{-7}
465	7.3900×10^{-2}	585	0.81630	705	2.9290×10^{-3}	825	6.4153×10^{-7}
470	9.0980×10^{-2}	590	0.75700	710	2.0910×10^{-3}	830	4.5181×10^{-7}
475	0.11260	595	0.69490	715	1.4840×10^{-3}		

Table 2.4. Definitive values of the spectral luminous efficiency function for scotopic vision $V'(\lambda)$

λ [nm]	V′(λ) [a.u.]	λ [nm]	V′(λ) [a.u.]	λ [nm]	V′(λ) [a.u.]	λ [nm]	V′(λ) [a.u.]
380	5.890×10^{-4}	485	8.510×10^{-1}	590	6.550×10^{-2}	695	2.501×10^{-5}
385	1.108×10^{-3}	490	9.040×10^{-1}	595	4.690×10^{-2}	700	1.780×10^{-5}
390	2.209×10^{-3}	495	9.490×10^{-1}	600	3.315×10^{-2}	705	1.273×10^{-5}
395	4.530×10^{-3}	500	9.820×10^{-1}	605	2.312×10^{-2}	710	9.140×10^{-6}
400	9.290×10^{-3}	505	9.980×10^{-1}	610	1.593×10^{-2}	715	6.600×10^{-6}
405	1.852×10^{-2}	510	9.970×10^{-1}	615	1.088×10^{-2}	720	4.780×10^{-6}
410	3.484×10^{-2}	515	9.750×10^{-1}	620	7.370×10^{-3}	725	3.482×10^{-6}
415	6.040×10^{-2}	520	9.350×10^{-1}	625	4.970×10^{-3}	730	2.546×10^{-6}
420	9.660×10^{-2}	525	8.800×10^{-1}	630	3.335×10^{-3}	735	1.870×10^{-6}
425	1.436×10^{-1}	530	8.110×10^{-1}	635	2.235×10^{-3}	740	1.379×10^{-6}
430	1.998×10^{-1}	535	7.330×10^{-1}	640	1.497×10^{-3}	745	1.022×10^{-6}
435	2.625×10^{-1}	540	6.500×10^{-1}	645	1.005×10^{-3}	750	7.600×10^{-7}
440	3.281×10^{-1}	545	5.640×10^{-1}	650	6.770×10^{-4}	755	5.670×10^{-7}
445	3.931×10^{-1}	550	4.810×10^{-1}	655	4.590×10^{-4}	760	4.250×10^{-7}
450	4.550×10^{-1}	555	4.020×10^{-1}	660	3.129×10^{-4}	765	3.196×10^{-7}
455	5.130e×10^{-1}	560	3.288×10^{-1}	665	2.146×10^{-4}	770	2.413×10^{-7}
460	5.670×10^{-1}	565	2.639×10^{-1}	670	1.480×10^{-4}	775	1.829×10^{-7}
465	6.200×10^{-1}	570	2.076×10^{-1}	675	1.026×10^{-4}	780	1.390×10^{-7}
470	6.760×10^{-1}	575	1.602×10^{-1}	680	7.150×10^{-5}		
475	7.340×10^{-1}	580	1.212×10^{-1}	685	5.010×10^{-5}		
480	7.930×10^{-1}	585	8.990×10^{-2}	690	3.533×10^{-5}		

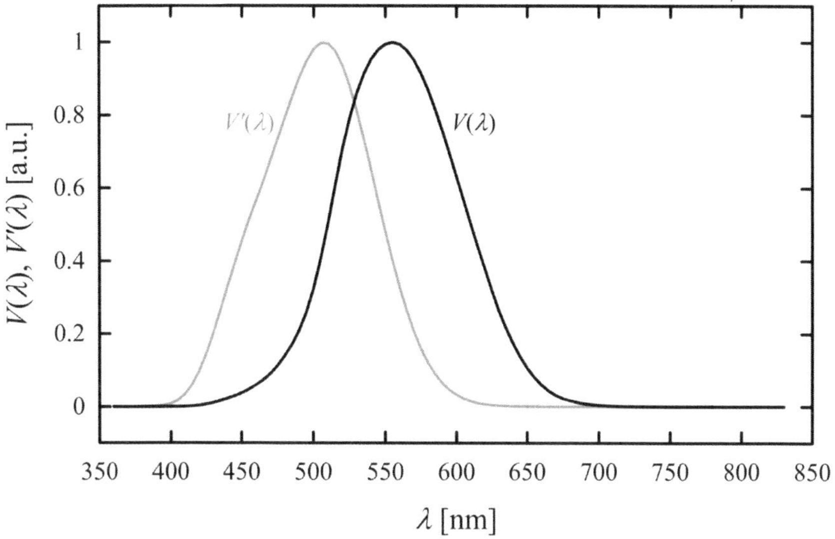

Figure 2.16 Spectral luminous efficiency function for photopic vision $V(\lambda)$ and for scotopic vision $V'(\lambda)$.

Standard Photometer and the Realization of the Candela

Typically, a standard photometer consists of a silicon photodiode, a $V(\lambda)$ correction filter and a photometer head with the precisely defined aperture opening, as presented in figure 2.17. If $s(\lambda)$ is the absolute spectral power responsivity of the whole photometer then the illuminance responsivity s_v of the photometer is given by:

$$s_V = \frac{A\int_\lambda S(\lambda)s(\lambda)d\lambda}{K_m \int_\lambda S(\lambda)V(\lambda)d\lambda}, \quad (2.34)$$

where A is the photometer head area of aperture opening and $S(\lambda)$ is the relative power spectral distribution of the light source to be measured. If the light source with the power spectral distribution of $S(\lambda)$ is measured with the photometer at a distance d, the illuminance at the photometer aperture is given by:

$$E_V = \frac{i_P}{s_V}, \quad (2.35)$$

where i_P is the of the photometer output current. If the distance d, measured from the light source to the photometer, is accurately known, then the luminous intensity of the source is given by:

$$I_V = \frac{E_V d^2}{\Omega_0} = \frac{i_P d^2}{s_V \Omega_0}, \qquad (2.36)$$

where $\Omega_0 = 1$ sr is the unit solid angle.

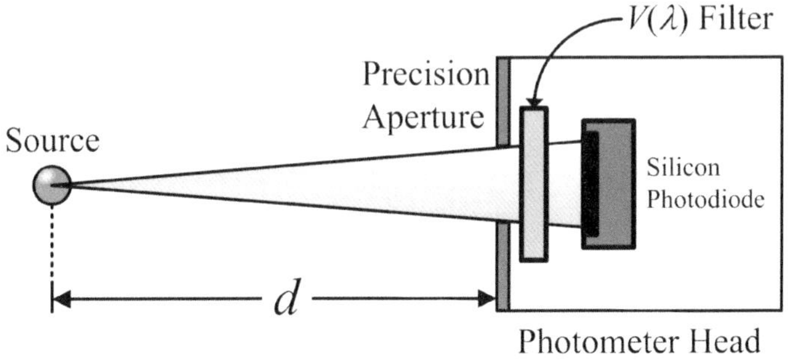

Figure 2.17. Detector-based candela realization.

REFERENCES

1. Brian D. Warner. *A Practical Guide to Lightcurve Photometry and Analysis*. New York: Springer Science+Business Media, Inc., 2006.
2. Michael Bukshtab. *Applied Photometry, Radiometry, and Measurements of Optical Losses*. Dordrecht: Springer Science + Business Media B.V., 2012.
3. E.F. Milone and C. Sterken (eds). *Astronomical Photometry: Past, Present, and Future*. New York: Springer Science+Business Media, LLC, 2011.
4. International Commission on Illumination. *The Basis of Physical Photometry*. Publication CIE No 18.2 (TC-I .2) 1983 Commission Internationale De L'éclairage 52, Boulevard Malesherbes - 75008 Paris – France.
5. Cornelius J. Willers. *Electro-Optical System Analysis and Design: A Radiometry Perspective*. Bellingham: SPIE – The International Society for Optical Engineering, 2013.

6. Barbara G. Grant. *Field Guide to Radiometry*. Bellingham: Society of Photo-Optical Instrumentation Engineers (SPIE), 2011.
7. Virendra N. Mahajan. *Fundamentals of Geometrical Optics*. Bellingham: Society of Photo-Optical Instrumentation Engineers (SPIE), 2014.
8. Edwin Budding and Osman Demircan. *Introduction to Astronomical Photometry (2nd Edition)*. Cambridge: Cambridge University Press, 2007.
9. William L. Wolfe. *Introduction to Radiometry*. Bellingham: The Society of Photo-Optical Instrumentation Engineers, 1998.
10. Robert Celotta and Thomas Lucatorto (eds). *Experimental Methods in the Physical Sciences: Volume 41 Optical Radiometry*. Amsterdam: Elsevier Inc., 2005.
11. Thomas Lucatorto and Albert C. Parr (eds). *Experimental Methods in the Physical Sciences: Volume 43 Radiometric Temperature Measurements*. Amsterdam: Elsevier Inc., 2010.

3
CHAPTER

GEOMETRICAL OPTICS AND OPTICAL INSTRUMENTS

LIGHT WAVE PROPAGATION AND OPTICAL RAYS

As shown in Chapter 1. one of the solution of the Maxwell's equations in the case when there are no electric charges and no electric currents, such as in a vacuum, is the plane wave. The plane wave is characterized by a one dimensional wave equation, where the wave amplitude is determined by the position along the axis of wave propagation:

$$U(z,t) = U_0 \cos(kz - \omega t), \qquad (3.1)$$

where U_0 is the corresponding field amplitude and where wave number k and angular frequency ω. Such a wave can be represented as a set of planes in a three dimensional space where the corresponding wave value has the constant value, i.e. the equiphase surfaces. Such an equiphase surface is normal to the wave vector, as shown in figure 3.1.

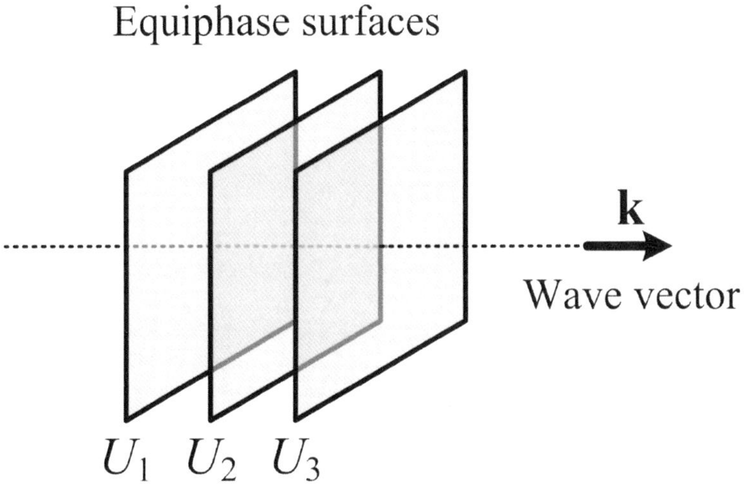

Figure 3.1. A set of three equiphase surfaces U_1, U_2, and U_3 and the corresponding wave vector **k**.

Bearing in mind that the wavelength of an electromagnetic radiation in the optical spectrum domain is very short (typically in the range of 1 μm) when compared with the dimensions of typical optical components such as mirrors, lenses, etc., a set of equiphase surfaces can be modeled and efficiently replaced by an optical ray that is collinear to the wave vector. An optical ray is used to model the propagation of light throughout an optical system, where the real light field is divided into the discrete rays, which can be easily mathematically processed in order to compute the way of their propagation throughout the optical system by the ray tracing techniques. In this way a very complex optical systems can be mathematically analyzed or simulated. Although the approximation of a real optical field with the optical rays is a powerful method in analyzing complex optical systems, it fails in the case when other wave phenomena such as diffraction and interference occur.

Optical Rays and Fermat's Principle

Each optical system consists of at least a light source that generates the optical rays and of the optical components where the reflection and/or refraction phenomena occur. Optical component deflects the rays making the different paths of the ray's propagation throughout the optical system thus forming the different ray traces. Depending on the optical system layout, the rays will follow the traces that are in agreement with the Fermat's principle. Fermat's principle or the principle of least time, named after French mathematician Pierre de Fermat in 1657, is the principle which states that the path taken between two points by an optical ray is the path that can be transited in the least time. This principle is sometimes taken as the definition of an optical ray. If we observe a light ray that propagates through the optical system along the path from the point A to the point B, as presented in figure 3.2., the Fermat's principle can be mathematically formulated as:

$$\tau = \int_A^B \frac{dl}{v} = \frac{1}{c}\int_A^B \frac{c}{v}dl = \frac{1}{c}\int_A^B n dl \quad , \qquad (3.2)$$

where the light rays follow the path between the points A and B where the time delay τ takes its minimal value, i.e.:

$$\text{ä}\tau = \frac{1}{c}\text{ä}\int_A^B n dl = 0 \quad . \qquad (3.3)$$

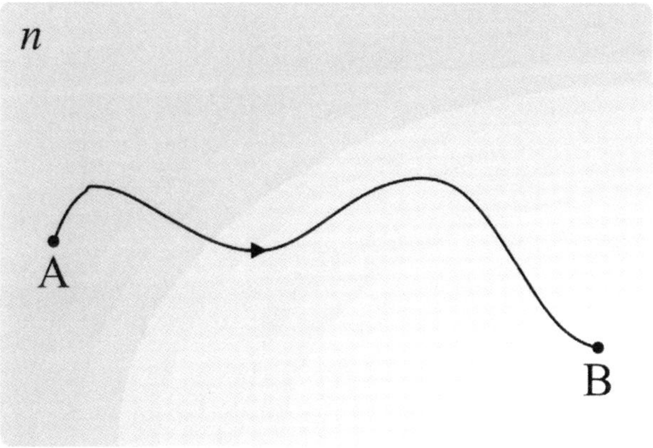

Figure 3.2. Fermat's principle definition.

By following the Fermat's principle one can derive two basic optical laws regarding the propagation of light rays, i.e. the law of reflection and the law of refraction. If we observe a light ray, as presented in figure 3.3, where the ray starts from point A and after the reflection from the mirror at point C arrives at point B, travels the overall path:

$$L = AC + CB = l_1 + l_2. \qquad (3.4)$$

By taking into account the geometrical relations between corresponding lengths we have:

$$L = \sqrt{h_1^2 + (d - x_2)^2} + \sqrt{h_2^2 + x_2^2}. \qquad (3.5)$$

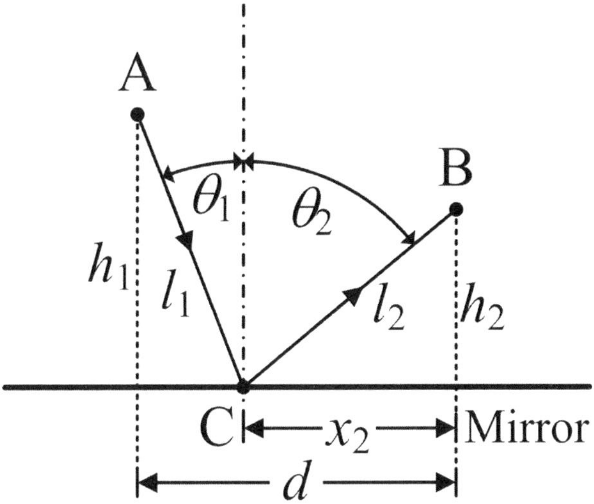

Figure 3.3. Reflection of ray.

Time of ray transit between points A and B is given by:

$$\tau = \frac{L}{c} = \frac{1}{c}\left[\sqrt{h_1^2 + (d - x_2)^2} + \sqrt{h_2^2 + x_2^2}\right]. \qquad (3.6)$$

According to the Fermat's principle the reflection at the point C at the mirror surface will occur in such a way that time duration, given by equation (3.6), takes the minimal value. The minimal value of transit time has been achieved when it is fulfilled:

$$\frac{d - x_2}{h_1} = \frac{x_2}{h_2}. \qquad (3.7)$$

Baring in mind that $\tan\theta_1 = (d - x_2)/h_1$ and $\tan\theta_2 = x_2/h_2$ we have:

$$\tan\theta_1 = \tan\theta_2 \quad \Rightarrow \quad \theta_1 = \theta_2, \qquad (3.8)$$

which represents the law of reflection.

Similarly, if we observe a light ray, as presented in figure 3.4., where the ray starts from point A and after the refraction at point C at the interface of two dielectric materials arrives at point B, travels the overall time:

$$\tau = \frac{\overline{AC}}{v_1} + \frac{\overline{CB}}{v_2} = \frac{l_1}{v_1} + \frac{l_2}{v_2}, \qquad (3.9)$$

where $v_1 = c/n_1$ and $v_2 = c/n_2$ are the corresponding speed of light in the dielectric medium. By taking into account the geometrical relations between corresponding lengths we have:

$$\tau = \frac{1}{c}\left[n_1\sqrt{h_1^2 + (d - x_2)^2} + n_2\sqrt{h_2^2 + x_2^2}\right]. \qquad (3.10)$$

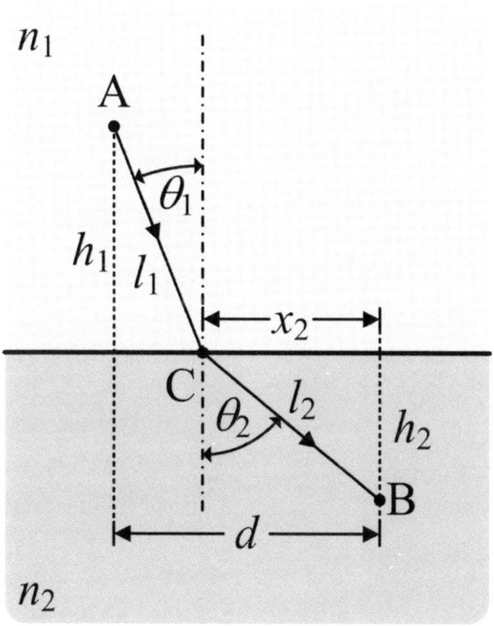

Figure 3.4. Refraction of ray.

According to the Fermat's principle the refraction at the point C at the interface of two dielectric materials will occur in such a way that time duration, given by equation (3.10), takes the minimal value. The minimal value of transit time has been achieved when it is fulfilled:

$$n_1 \frac{d - x_2}{l_1} = n_2 \frac{x_2}{l_2}. \qquad (3.11)$$

Baring in mind that $\sin\theta_1 = (d - x_2)/l_1$ and $\sin\theta_2 = x_2/l_2$ we have:

$$n_1 \sin\theta_1 = n_2 \sin\theta_2, \qquad (3.12)$$

which represents the law of refraction or the Snell's law. The same results for the laws of reflection and refraction can be derived if we observe the electromagnetic wave propagation at the perfectly conducting surface (mirror) and at the interface between two dielectric mediums taking into consideration the boundary conditions. One can noticed that in the case of the ray refraction, the angle between the axis and the ray is always smaller in the optically denser medium (i.e. the medium with larger index of refraction). Therefore, when the ray arrives from the optically denser medium to the optically less dense medium at a certain angle θ, where it is fulfilled, $n_1 \sin\theta \geq n_2$ $(n_1 > n_2)$ there are no refraction and total internal reflection occurs. The critical angle is given by:

$$\theta_C = \arcsin\left(\frac{n_2}{n_1}\right), \qquad (3.13)$$

where for the angles greater than the critical angle the total internal reflection occurs. One of the interesting ways of the total internal reflection application is the fingerprint capturing. The process of fingerprint capturing based on the total internal reflection is presented in figure 3.5. In order to capture the shape of the finger ridges a finger is placed onto the flat surface of the optical prism. The finger is illuminated by the light source that emits almost parallel rays. Depending does the light ray impinge the finger ridge or valley, the ray is totally internally reflected when the ray impinges the finger valley. In the case when the ray hits the ridges, the light is scattered at the finger surface and thus not captured by the camera. Therefore, an image of the fingerprint valleys is captured by the camera. In order to get the image of the ridges and to make image compliant with older fingerprint capturing methods a negative of the captured image must be made.

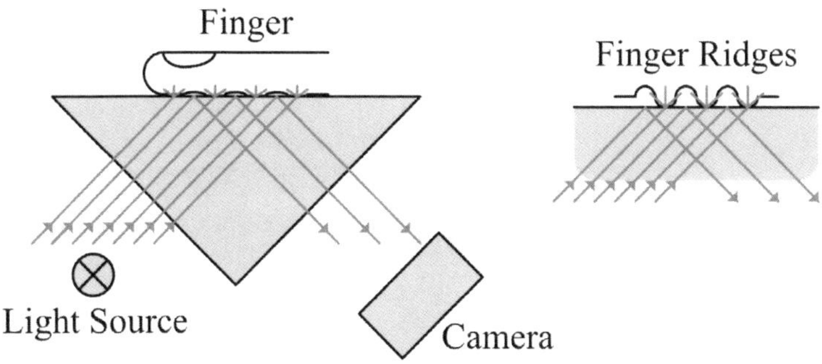

Figure 3.5. Fingerprint capturing process by the total internal reflection.

Plane-Parallel Plates and Prisms

A plane-parallel plate is typically a glass plate which is the most common optical component that can be found in many applications such as beam splitters, cover glasses, optical windows, optical filters, etc. When the plane-parallel plate is placed along the light ray path, the lateral ray shift occurs, as presented in figure 3.6. Based on the geometrical relations and the law of refraction the lateral shift *d* is given by a well-known formula:

$$\delta = t\left(1 - \frac{\cos\alpha}{\sqrt{n^2 - \sin^2\alpha}}\right)\sin\alpha \quad , \qquad (3.14)$$

where t is the plate thickness, n is the refractive index and α is the angle between the ray and the plate. In the most common case when we have $n = 1.5$ and $\alpha = 45°$ the lateral shift is $\delta \approx t/3$.

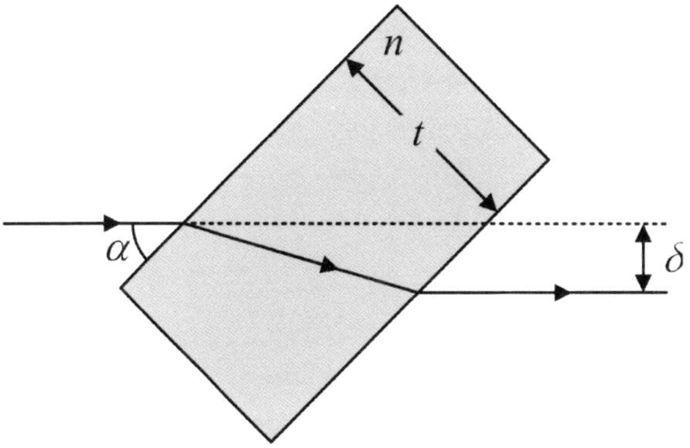

Figure 3.6. Lateral ray shift on the plane-parallel glass plate.

Important thing about the glass plates is that they introduce spherical and chromatic aberration when installed in focused beam as presented in figure 3.7. For the paraxial focusing. i.e. when the rays are almost parallel to the main optical axis the longitudinal displacement is given by:

$$\Delta = t\left(1-\frac{1}{n}\right), \qquad (3.15)$$

where also in the most common case, when we have $n=1.5$, the longitudinal shift is $\Delta \approx t/3$. Lateral and longitudinal displacements of the rays, given by equations (3.14) and (3.15), are well known. However, the lateral and longitudinal displacements of the light rays occur simultaneously when the inclined glass plate has been positioned into the focusing beam as presented in figure 3.8. Such optical setup can be often found when the beam splitter in the shape of the glass plate has been used. Beside the reflected rays, which are not important for the analysis, there are focused rays that pass thorough the beam splitter. These rays suffer from both lateral and longitudinal displacements. In order to calculate both displacements a closer look of the figure 3.8. with all relevant geometric parameters has been presented in figure 3.9.

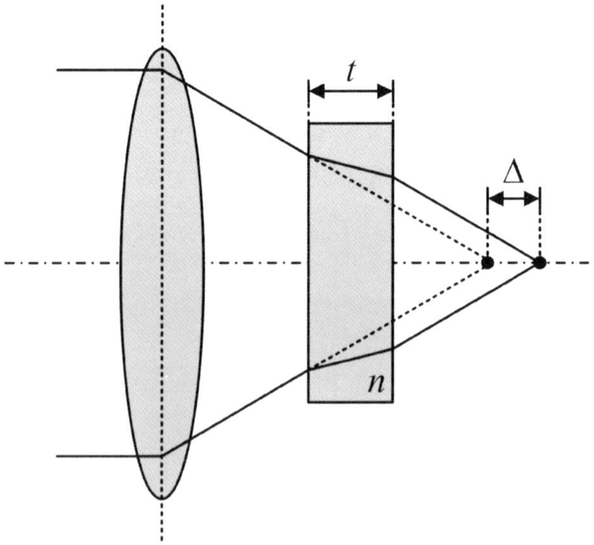

Figure 3.7. Longitudinal ray shift of the focused beam.

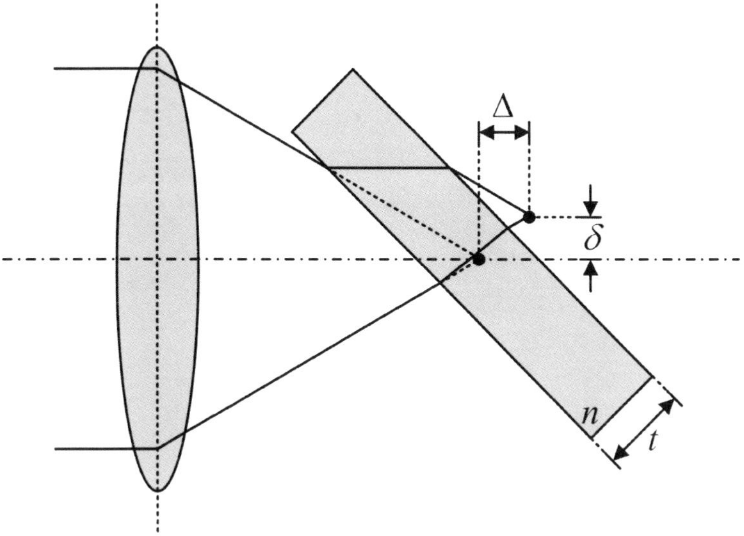

Figure 3.8. Longitudinal and lateral ray shift of the focused beam.

Based on the Snell's law we can write:

$$\sin\beta = \frac{\sin\alpha}{n}, \quad \sin(\beta+\Delta\beta) = \frac{\sin(\alpha+\Delta\alpha)}{n}, \tag{3.16}$$

where n is the plate refractive index, α, $\alpha+\Delta\alpha$, β, and $\beta+\Delta\beta$ are the angles presented in figure 3.9. where for the paraxial rays is valid $\Delta\alpha << 1$, $\Delta\beta << 1$. After rearranging equation (3.16) becomes:

$$\Delta\beta \approx \frac{\cos\alpha}{n\cos\beta}\Delta\alpha \,. \tag{3.17}$$

Geometry from figure 3.9. gives the following relations:

$$\mathrm{AB} = \frac{t}{\cos(\beta+\Delta\beta)}, \quad \mathrm{AB}' = \mathrm{AB}\cos[\alpha-(\beta+\Delta\beta)], \quad \mathrm{BB}' = \mathrm{AB}\sin[\alpha-(\beta+\Delta\beta)] \tag{3.18}$$

and

$$\mathrm{B} = \mathrm{AA}' + \mathrm{BB}', \quad \mathrm{BD}' = \mathrm{B} - \delta \tag{3.19}$$

and also

$$\Delta = \mathrm{AB}' + \mathrm{D'F'} - \mathrm{A'F}, \quad \mathrm{D'F'} \approx \frac{\mathrm{BD}'}{\Delta\alpha}, \quad \mathrm{A'F} \approx \frac{\mathrm{AA}'}{\Delta\alpha} \,. \tag{3.20}$$

Equation (3.18) gives:

$$\mathrm{AB}' = t\frac{\cos[\alpha-(\beta+\Delta\beta)]}{\cos(\beta+\Delta\beta)}, \quad \mathrm{BB}' = t\frac{\sin[\alpha-(\beta+\Delta\beta)]}{\cos(\beta+\Delta\beta)} \quad . \tag{3.21}$$

By combining equations (3.19) and (3.21) we have:

$$\mathrm{BD}' = \mathrm{AA}' + t\frac{\sin[\alpha-(\beta+\Delta\beta)]}{\cos(\beta+\Delta\beta)} - \delta \quad . \tag{3.22}$$

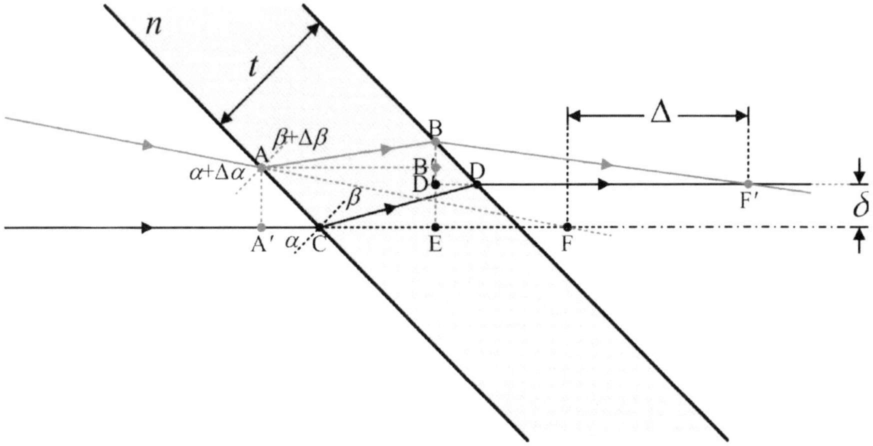

Figure 3.9. Closer look to the longitudinal and lateral ray shift of the focused beam.

Further, the combination of the last three equations gives:

$$\Delta \approx t\frac{\cos[\alpha-(\beta+\Delta\beta)]}{\cos(\beta+\Delta\beta)} + \frac{t}{\Delta\alpha}\frac{\sin[\alpha-(\beta+\Delta\beta)]}{\cos(\beta+\Delta\beta)} - \frac{\delta}{\Delta\alpha} \quad . \tag{3.23}$$

Finally, for the longitudinal displacement we have:

$$\Delta \approx t\left[\cos\alpha + \frac{\sin^2\alpha}{\sqrt{n^2-\sin^2\alpha}} - \frac{n^2\cos^2\alpha}{\sqrt{\left(n^2-\sin^2\alpha\right)^3}}\right] \tag{3.24}$$

and for the lateral displacement obviously we have the same value as given by equation (3.14). By taking into account that typically we have $n = 1.5$ and $\alpha = 45°$, the longitudinal shift is $\Delta = 0.6t$.

When a light ray impinges an optical prism, a wedge-shaped piece of glass, the ray is refracted twice, as presented in figure 3.10. Due to the double refraction the ray suffers from the deviation, which angle of deviation δ depends on the prism angle α, incident angle β, and its index of refraction n in the following way:

$$\delta = \alpha - \beta + \arcsin\left(\cos\alpha\sin\beta - \sin\alpha\sqrt{n^2 - \sin^2\alpha}\right). \quad (3.25)$$

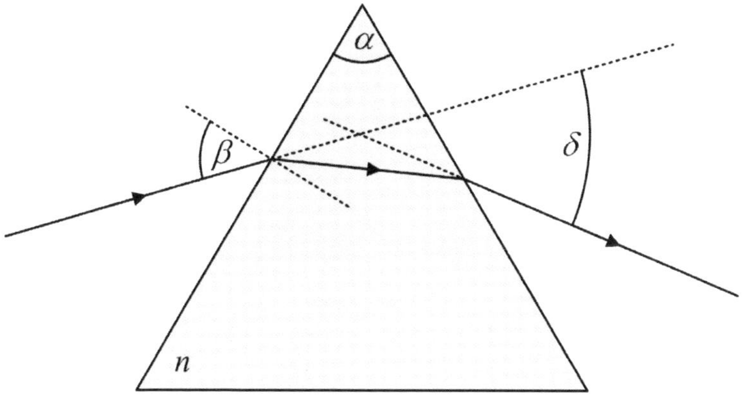

Figure 3.10. Light ray deflection by prism.

If we plot the deviation angle versus the incident angles, there is a range of deviation angles that are impossible to obtain. Figure 3.11. represents a plot of prism deviation angle versus incident angle and prism angle with a fixed index of refraction. One can notice that there is a minimum deviation angle which depends on a index of refraction and a prism apex angle. By changing the incident angle there is a possibility to measure the prism refractive index simply by measuring the minimal deviation angle. Minimal deviation angle can be found from equation (3.25) by equating its first derivative with respect to incident angle and finding the root of such an equation. The minimal deviation angle is given by:

$$\delta_{\min} = \alpha - 2\arcsin\left[n\sin\left(\frac{\alpha}{2}\right)\right]. \quad (3.26)$$

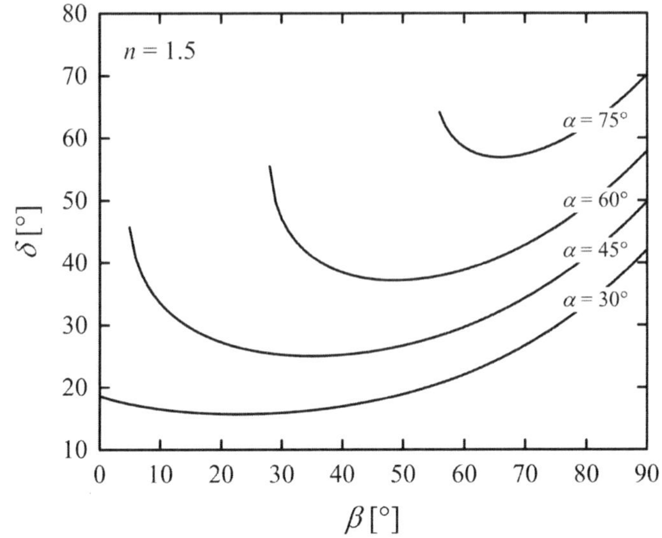

Figure 3.11. Light ray deflection by prism versus incident angle and prism apex angle.

If the angle of incidence on the first surface is equal to the angle of incidence on the second surface the minimum deviation angle is obtained. In this way the ray within the prism intersect the prism apex angle bisection line at a right angle. By finding the minimum deviation angle, from equation (3.27) one can obtain the prism index of refraction in the following way:

$$n = \frac{\sin\left(\frac{\alpha - \delta_{min}}{2}\right)}{\sin\left(\frac{\alpha}{2}\right)} \qquad (3.27)$$

If the prism apex angle is small ($\sin\alpha \approx \alpha$), like in the case of an optical wedge, equation (3.26) becomes:

$$\delta_{min} \approx \alpha(n-1) \qquad (3.27)$$

The total internal reflection, which can occur at inner prism walls, is used in the deviating prisms in order to correct the handedness of an image and to change the direction of the ray. Beside the lens the deviating prism is most commonly used optical component. The typical application of a deviation prism is in binoculars, where it used to make the invert image (upright image). Deviating prisms can be classified according to their deviation angle. The various types of designs are presented in figure 3.12. for (a) 45°, (b) 90° and (c) 180° deviating prisms.

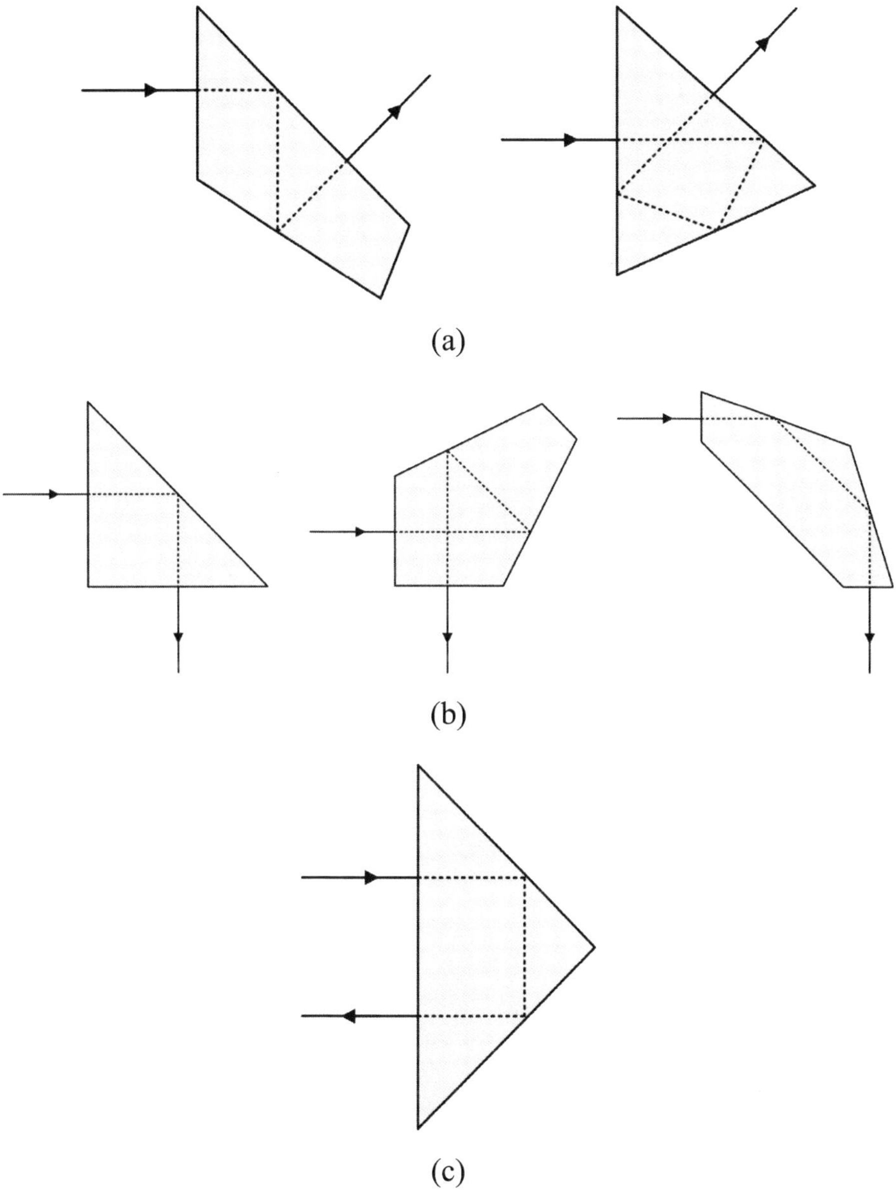

Figure 3.12. Deviating prisms for (a) 45°, (b) 90° (right angle, penta, and Wollaston prism), and (c) 180° deviation angle.

Maybe the most outstanding applications of deviating prisms is in the high resolution and high performance color cameras such one used in the recording studios. Typical optical layout of such a camera is presented in figure 3.13. where the dichroic color filters F_1 and F_2 were used between the prisms in order to separate three color channels red (R), green (G), and blue (B). After the separation each channel is separately captured by the *CCD (Charge-Coupled* ***D****evice)* camera.

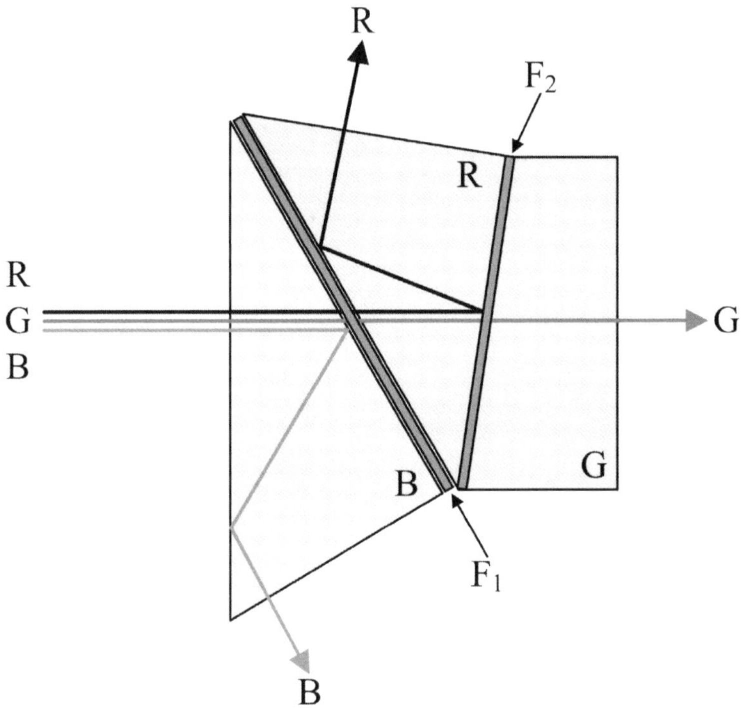

Figure 3.13. Deviating prisms in high performance color camera.

As mentioned above the deviation angle of a prism depends not only on the prism geometry but also on the index of refraction of a material the prism has been made of. However, the index of refraction of any material depends on the used wavelength of a light. According to Sellmeier formula the index of refraction depends on a wavelength in the following way:

$$n^2(\lambda) = 1 + \sum_{i=1}^{3} \frac{a_i \lambda^2}{\lambda^2 - b_i}, \qquad (3.28)$$

where a_i and b_i are material dependent Sellmeier coefficients. Sellmeier formula is widely used in optical science and optical industry to describe and characterize the dispersion in glasses and crystals. In the case of a commonly used borosilicate crown glass known as BK7 the Sellmeier coefficients are shown in table 3.1.

Table 3.1. Sellmeier coefficients for borosilicate crown glass

Coefficient	Value
a_1	1.03961212
a_2	0.231792344
a_3	1.01046945
b_1	6.00069867×10^{-3} µm^2
b_2	2.00179144×10^{-2} µm^2
b_3	1.03560653×10^{2} µm^2

Due to prism glass dispersion, the light, which is composed of different wavelengths, will bent differently for different wavelengths. According to Sellmeier formula the refractive index of the prism is larger for blue then for red light. Therefore, based on the result of equation (3.27), the blue light will bent more than the red light thus separating spatially each spectral component of the incoming light, as it is presented in figure 3.14.

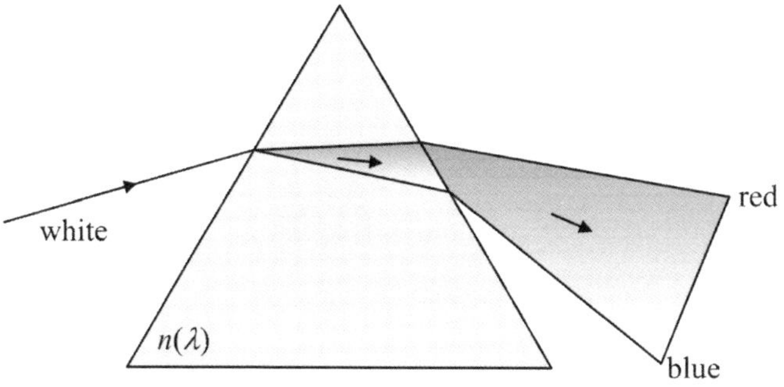

Figure 3.14. Dispersing prism and spectral decomposition.

The dispersion in a prism can be used affirmatively in order to make for example an optical spectrometer. However, the dispersion can be a big issue if a short light pulse must be transferred though the dispersive material such an optical fiber. Optical fibers are most commonly used medium for information transport all over the world. The every growing demand for a larger capacity pushes the use of a very short light pulses in order to provide high capacity. One of the limiting factors is surely the fiber chromatic dispersion as it stretches the light pulse that propagates within the optical fiber. For example, if we observe an optical fiber with the length *l*, as presented in figure 3.15., the width of an optical pulse will be in time stretched to the pulse width of τ. The light with shorter wavelength (for example blue light) will lead the stretched pulse, where the light with the longer wavelength light (for example red light) will be on the pulse tail end.

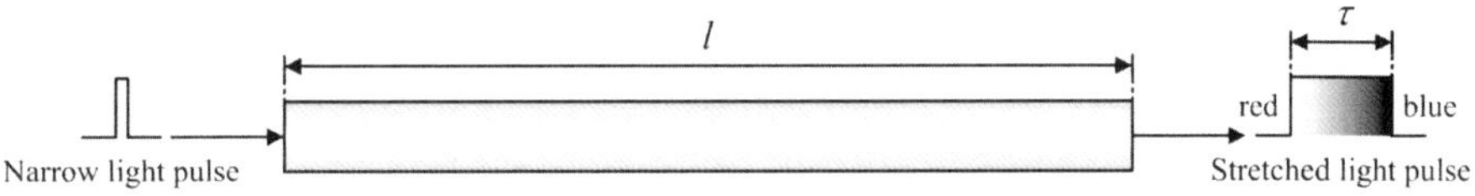

Figure 3.15. Chromatic dispersion in an optical fiber.

If the refractive index difference of the optical fiber in the spectral range of the transferring light pulse is Δn, then the light pulse width at the end of the optical fiber will be given with:

$$\tau = \frac{\Delta n}{c} l \quad , \qquad (3.29)$$

where the refractive index difference depends on the optical fiber glass type and the spectrum width of the light, thus limiting the capacity of an optical fiber to the value of:

$$C\left[\frac{\text{bits}}{\text{s}}\right] = \frac{1}{\tau} = \frac{c}{\Delta h} \quad . \qquad (3.30)$$

The Abbe number is a quantitative measure of the dispersion curve average slope i.e. the average slope of the refractive index versus wavelength curve, as presented in figure 3.16. This slope represents in fact the dispersive characteristics of the optical medium. If the curve is flat, i.e. if the index of refraction doesn't change its value with the wavelength the optical medium does not disperse. The Abbe number, also known as the V-number or constringence, characterizes the glass dispersion as:

$$V^{\#} = \frac{n_d - 1}{n_F - n_C}, \quad (3.31)$$

where n_F, n_d and n_C are the index of refraction values of a glass for defined wavelengths, as presented in table 3.2.

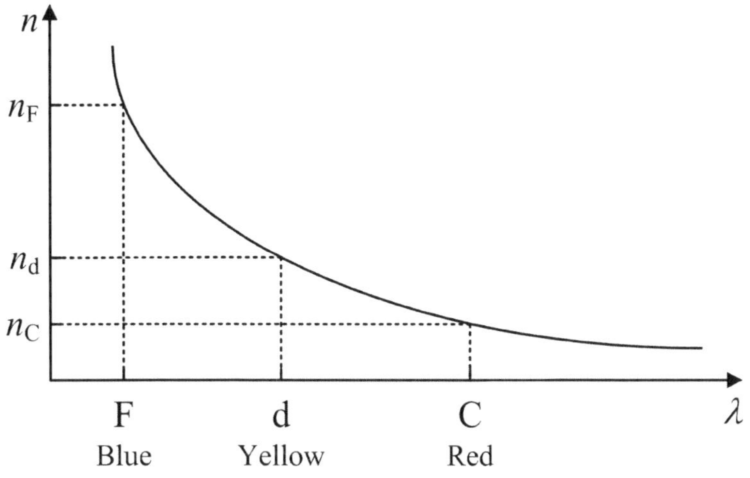

Figure 3.16. Typical dispersion curve.

If the refractive index difference at the F and the C wavelengths is a small, the Abbe number is large, which indicates a small dispersion. In opposite, a large dispersion glass has a low Abbe number. Typical range of the Abbe number values is from about 20 to 90. A glass which has Abbe numbers greater than 55 are classified as crown glasses, and glasses with Abbe numbers less than 50 are classified as flint glasses.

Table 3.2. Index of refraction values of a glass for defined wavelengths

Index	Color	Frequency [Hz]	Wave-length [nm]
F	Blue	6.172×10^{14}	486.1
d	Yellow	5.106×10^{14}	587.6
C	Red	4.570×10^{14}	656.3

As previously mentioned due to the glass dispersion the deviation angle of a prism will differ for the different wavelengths. According to equation (3.27), the minimal spread in prism deviation angle can be found as the minimal difference in the ray deviation, as presented in figure 3.17, and it will be given as

$$\Delta\delta_{\mathrm{min}} \approx \alpha\Delta n = \frac{\delta_{\mathrm{min}}}{n-1}\Delta n \quad , \tag{3.32}$$

where Δn is the refractive index variation which can be found as:

$$\Delta\delta_{\mathrm{min}} \approx \delta_{\mathrm{min}} \frac{n_{\mathrm{d}}-1}{n_{\mathrm{F}}-n_{\mathrm{C}}} = \delta_{\mathrm{min}} V^{\#} \quad . \tag{3.33}$$

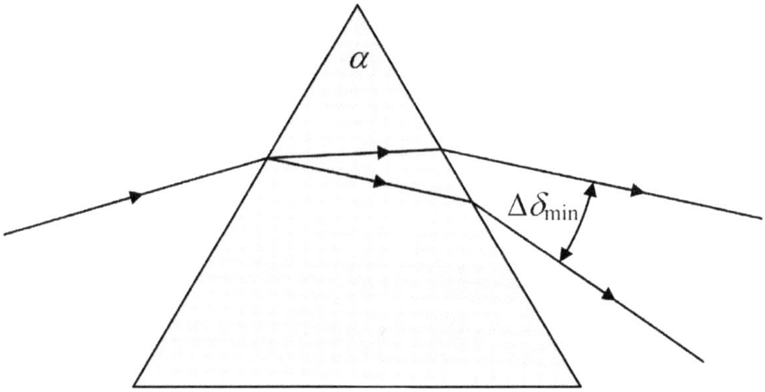

Figure 3.17. Ray deviation difference.

If we combine two different types of prisms we can expect two things. The first one is to get deviation without the dispersion and the second one is to get dispersion without deviation. If we use a pair of prisms that consist of crown and flint glass prisms in the case when we do not have dispersion for C and F wavelengths, as presented in figure 3.18 (a), the following condition must be fulfilled:

$$\begin{gathered} \delta_{\mathrm{C}}^{\mathrm{Crown}} + \delta_{\mathrm{C}}^{\mathrm{Flint}} = \delta_{\mathrm{F}}^{\mathrm{Crown}} = \delta_{\mathrm{F}}^{\mathrm{Flint}} \\ \delta_{\mathrm{d}} = \delta_{\mathrm{d}}^{\mathrm{Crown}} + \delta_{\mathrm{d}}^{\mathrm{Flint}} \quad , \end{gathered} \tag{3.34}$$

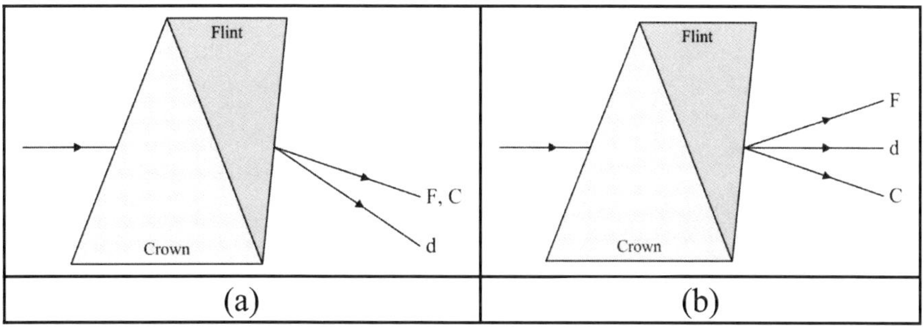

Figure 3.18. (a) Deviation without dispersion and (b) dispersion with no deviation.

In the case when d light has no deviation, as presented in figure 3.18 (b), the deviation of the crown prism is compensated by an equal but opposite deviation of the flint prism, i.e. the following must be fulfilled:

$$\delta_{\mathrm{d}}^{\mathrm{Crown}} = -\delta_{\mathrm{d}}^{\mathrm{Flint}} \; . \tag{3.35}$$

Spherical Mirrors

In order to maximally suppress the aberration of an imaging system the precise aspheric surfaces, such as parabolic or elliptical mirrors, must be used. However, the production of such mirrors are considerably more difficult than the fabrication of the spherical ones, and, consequently, they are considerably more expensive. Therefore, we will analyze in details the spherical mirror configurations in order to determine the circumstances under which it can replace the aspheric mirrors. The parabolic mirror cross section is defined by the well-known polynomial equation:

$$x = \frac{y^2}{4f} \; , \tag{3.36}$$

where f is the focal length of the parabolic mirror. On the other side, the corresponding spherical mirror cross section is given as:

$$\begin{aligned} (x-R)^2 + y^2 &= R^2 \\ x &= R \pm \sqrt{R^2 - y^2} \; . \end{aligned} \tag{3.37}$$

If we further develop equation (3.37) into the Maclaurin series we have:

$$x = \frac{y^2}{2R} + \frac{y^4}{8R^3} + O\left(y^6\right), \tag{3.38}$$

where $O(\bullet)$ represents the big o notation. By comparing equations (3.36) and (3.38) one can notice that spherical mirror can replace the parabolic mirror if it is satisfied:

$$f = \frac{R}{2}$$
$$y < \ 2R, \tag{3.39}$$

where the first relation of the last equation is the well-known focus length of the spherical mirror and the second on represents the paraxial condition. In the paraxial region, i.e. in the immediate vicinity of the mirror central axis, parabolic and spherical mirrors will be essentially indistinguishable. The comparison between the parabolic mirror and the corresponding equivalent spherical mirror is given in figure 3.19.

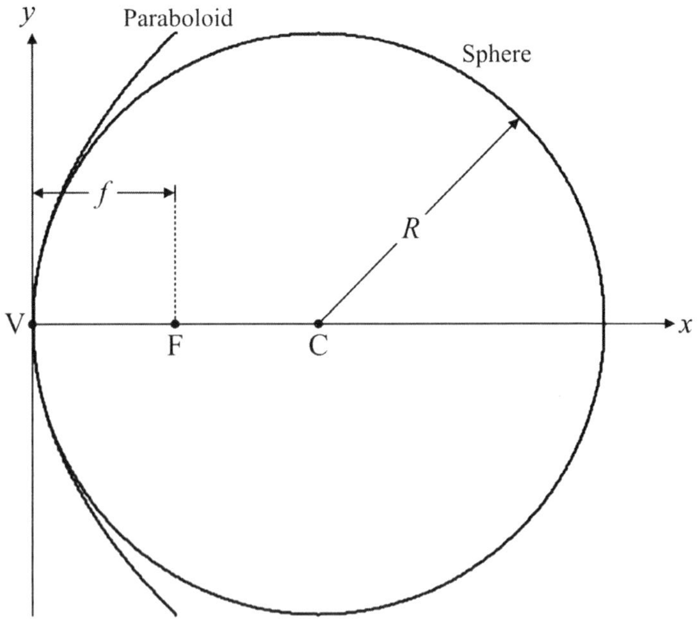

Figure 3.19. Equivalent parabolic and spherical mirrors.

Taking into consideration the paraxial theory of the spherical mirrors as a first approximation, one can find the paraxial equation of the spherical mirror that relates object and image positions along the optical axis according to the geometrical representation given in figure 3.20.

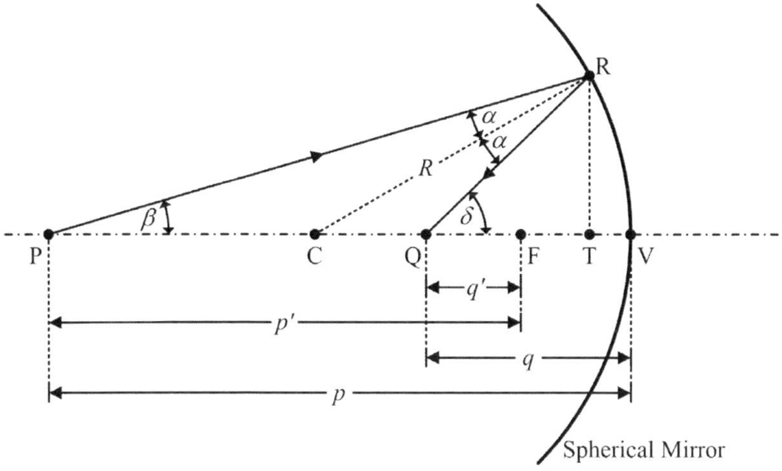

Figure 3.20. Concave spherical mirror.

For a small angle β (paraxial approximation β << 1) we have $R \approx R$ and consequently $V < P = p$ and $V < Q = q$. Therefore, we have:

$$R = p \tan\beta = q \tan\delta \quad , \qquad (3.40)$$

or in the paraxial approximation:

$$p\beta \approx q\delta \quad . \qquad (3.41)$$

From the triangle ΔPRQ we have $\delta = 2\alpha + \beta$, so we obtain:

$$\frac{p-q}{q} = \frac{2\alpha}{\beta} \quad . \qquad (3.42)$$

According to the low of sines from the triangle ΔCPR yields:

$$\frac{\sin\alpha}{\sin\beta} \approx \frac{\alpha}{\beta} = \frac{C}{R} = \frac{p-R}{R} \quad . \qquad (3.43)$$

By combining equations (3.42) and (3.43) we have:

$$\frac{p-q}{q} = 2\frac{p-R}{R} \quad , \qquad (3.44)$$

or finally, by rearranging equation (3.44) we obtain the well-known spherical mirror formula:

$$\frac{1}{p}+\frac{1}{q}=\frac{2}{R}=\frac{1}{f}. \qquad (3.45)$$

If we go back to figure 3.20. we can see that $q = f + q'$ and $p = f + p'$ is valid. By substituting these values into equation (3.45), it becomes:

$$p'q' = f^2, \qquad (3.46)$$

which is the well-known Newton's formula. The presented analysis was performed for the case of concave mirror. The same relation between the object and image distances and mirror focal length is valid in the case of a convex mirror but it must be taken into account that convex mirror has negative radius of curvature ($R < 0$) and consequently has negative focal length ($f < 0$).

The process of graphical image location is straightforward. Figure 3.21. represents a concave mirror image formation when the object is at the distance larger than the radius of spherical mirror curvature ($p > R$) with all relevant rays for the graphical determination of an image position.

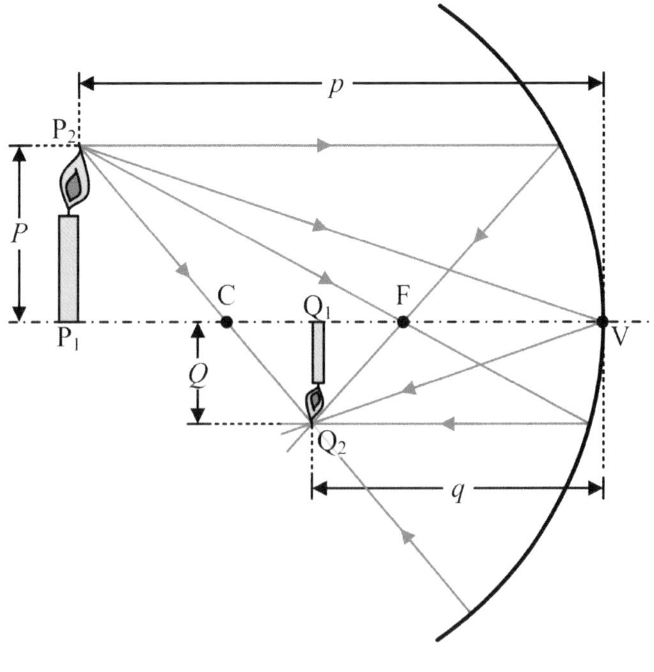

Figure 3.21. Concave mirror image formation.

As presented in figure 3.21. there are four relevant mirror rays and there are their main characteristics:

- The ray that intersect the center of curvature of a spherical mirror C is back reflected again through the center of curvature,
- The ray that intersect the focal point F is reflected parallel to the optical axis,
- The ray that impinge mirror vertex is reflected symmetrically to the optical axis, and
- The ray that propagate parallel to the optical axis is reflected through the focal point F.

The image position is determined by the equation (3.45). However, the size of an image can be determined by the observation of similar triangles ΔP_1P_2V and ΔQ_1Q_2V as follows:

$$\frac{P_1P_2}{P_1V} = \frac{Q_1Q_2}{Q_1V} . \qquad (3.47)$$

Having in mind that $P_1P_2 = P$, $P_1V = p$, $Q_1Q_2 = Q$, and $Q_1V = q$ is valid, the image magnification is given by:

$$M = \frac{Q}{P} = \frac{q}{p} . \qquad (3.48)$$

There is an interesting "optical illusion" made by two identical concave spherical mirrors facing each other as presented in figure 3.22. If a small object is placed at the bottom of the subjacent mirror M2 an image of the object will be show at the small opening positioned at the center of the upper mirror M1. So, virtually, an observer will see the object that hoovers in the opening.

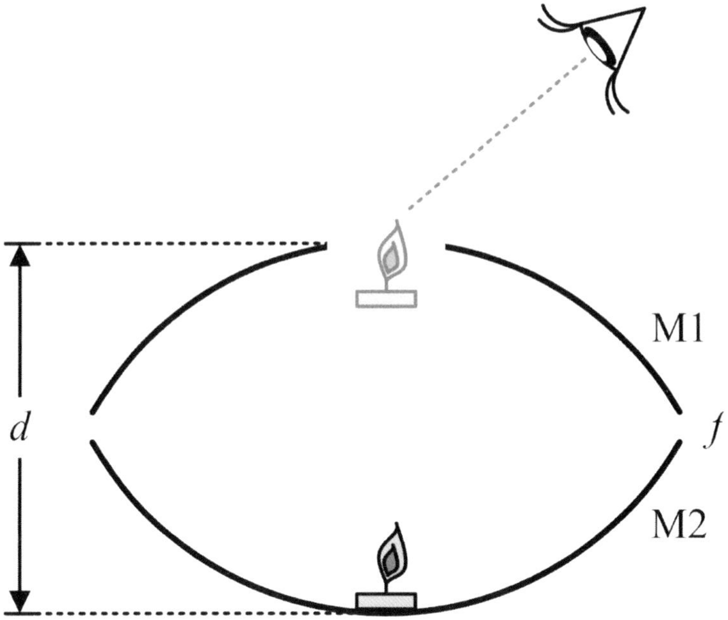

Figure 3.22. "Optical illusion" made by two concave spherical mirrors.

If the distance between the mirrors is d and the object is small if compared with the mirror dimensions, then the image that is formed by the upper mirror will be positioned at the distance:

$$q_1 = \frac{fd}{d-f} \quad , \qquad (3.49)$$

where f is the focal length of the mirrors (they are identical mirrors so both of them have the same focal length). The image of the upper mirror is at the same time the object of the subjacent mirror, where the position of the subjacent mirror object is given by:

$$p_2 = d - q_1 \, . \qquad (3.50)$$

The image of the subjacent mirror will be at the position:

$$q_2 = \frac{p_2 f}{p_2 - f} \, . \qquad (3.51)$$

In order to see object "hovering" the observer must fulfill the following condition:

$$q_2 = d\,. \tag{3.52}$$

The combination of the last four equations gives us:

$$d^2 - 4fd + 3f^2 = 0\,, \tag{3.53}$$

where the solutions of this quadratic equation is given by $d_{1,2} = (2 \pm 1)f$. Simply by choosing the shorter distance $d = f$ between the mirrors, i.e. by positioning the mirrors at the distance that is equal to their focal length, one can obtain object "hovering" optical illusion with the smaller dimensions of the overall setup. Moreover, the "hovering" object will be of the same dimensions as the original one as the presented optical setup has no magnification (magnification is equal to unity).

Refraction at Spherical Surfaces

The refraction at the spherical surfaces will be analyzed according to the optical setup presented in figure 3.23., where we have a semi-infinite optical slab that at its entrance has a spherical surface with the radius *R*. The slab, with the refractive index n_2, is immersed into another dielectric material with the index of refraction equal to n_1. In order to find the mathematical formula that relates all relevant parameters, we will observe the object point P at the optical axis as the source of optical rays and the particular ray that emanates from the point P at the angle θ_1 with respect to the optical axis.

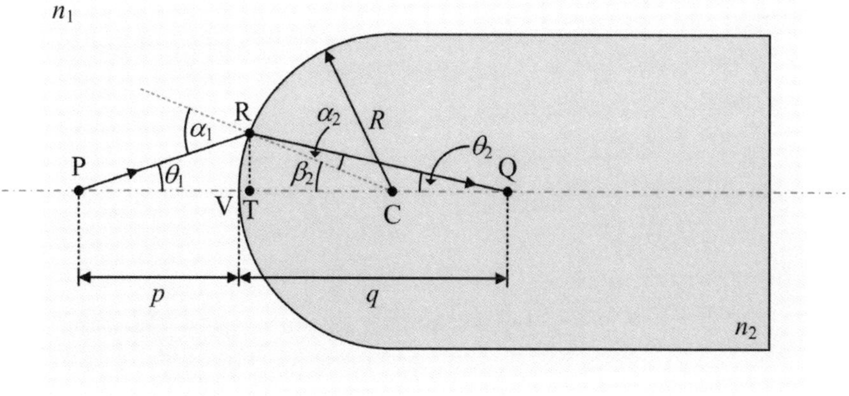

Figure 3.23. Refraction at spherical surface.

According to the Snell's law at the point of the refraction R we have:

$$n_1 \sin\alpha_1 = n_2 \sin\alpha_2, \qquad (3.54)$$

or in the case of paraxial rays:

$$n_1\alpha_1 \approx n_2\alpha_2. \qquad (3.55)$$

Based on the geometry presented in figure 3.23., we have:

$$R = P \sin\theta_1 = Q \sin\theta_2. \qquad (3.56)$$

Taking into consideration the paraxial approximation, the following is valid: $V < R$, $P \approx p$, $Q \approx q$, $\sin\theta_1 \approx \theta_1$, and $\sin\theta_2 \approx \theta_2$. In this case equation (3.56) becomes:

$$p\theta_1 \approx q\theta_2. \qquad (3.57)$$

Having in mind that for triangles ÄCQR and ÄCPR the following is valid:

$$\beta_2 = \alpha_2 + \theta_2$$
$$\alpha_1 = \theta_1 + \beta_2, \qquad (3.58)$$

and according to the low of sines from the triangle ÄCQR:

$$\frac{\sin\alpha_2}{Q} = \frac{\sin\theta_2}{R}, \qquad (3.59)$$

where due to the paraxial approximation we have:

$$\frac{\alpha_2}{\theta_2} = \frac{Q}{R} = \frac{q-R}{R}, \qquad (3.60)$$

we finally obtain from the last analysis the law of refraction at the spherical surface:

$$\frac{n_1}{p} + \frac{n_2}{q} = \frac{n_2 - n_1}{R}. \qquad (3.61)$$

Thin Lens

The above presented mathematical model aimed for the analysis of the spherical surface refraction, we will apply in order to determine the equation of a thin less presented in figure 3.24. A thin lens has two spherical surfaces and thus two refractions occur at the interface. For a thin lens the thickness is much smaller if compared with other lens dimensions, i.e. $t < R_1, R_2$ is valid (this condition is typically fulfilled for an eyeglass).

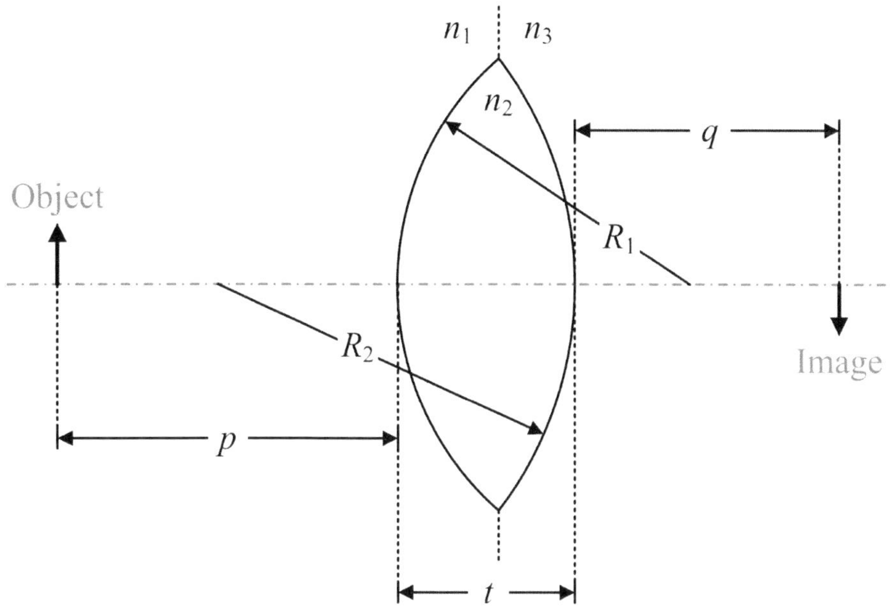

Figure 3.24. Thin convex lens.

If we apply the previously derived low of the spherical surface refraction, given by equation (3.61), at the first surface with the radius of curvature R_1 we obtain:

$$\frac{n_1}{p_1} + \frac{n_2}{q_1} = \frac{n_2 - n_1}{R_1}, \qquad (3.62)$$

where $p_1 = p$ is the object distance and q_1 is the image distance for the first surface. Similarly we have for the second surface:

$$\frac{n_2}{p_2} + \frac{n_3}{q_2} = \frac{n_3 - n_2}{R_2}, \qquad (3.63)$$

where p_2 is the object distance and $q_2 = q$ is the image distance for the second surface. Bearing in mind that $p_2 = -q_1$ is valid (the lens thickness is small so we omit it in the account), equation (3.62) and (3.63) give the following:

$$\frac{n_1}{p}+\frac{n_3}{q}=\frac{n_2-n_1}{R_1}+\frac{n_3-n_2}{R_2}\quad (3.64)$$

If a lens is immersed in the dielectric material then we have fulfilled $n_3 = n_1$ and equation (3.64) becomes:

$$\frac{1}{p}+\frac{1}{q}=\frac{n_2-n_1}{n_1}\left(\frac{1}{R_1}-\frac{1}{R_2}\right)\quad (3.65)$$

Therefore, the thin lens focal length f is given by:

$$\frac{1}{f}=\frac{n_2-n_1}{n_1}\left(\frac{1}{R_1}-\frac{1}{R_2}\right)\quad (3.66)$$

This last equation is called the *lensmaker's equation*. In most cases the surrounding medium is air with $n_1 = 1$, so equation (3.66) becomes:

$$\frac{1}{f}=(n-1)\left(\frac{1}{R_1}-\frac{1}{R_2}\right)\quad (3.67)$$

where $n_2 = n$ was taken into account. The process of graphical image location is straightforward. Figure 3.25. represents a thin convex lens formation when the object is at the distance larger than the focal length of the lens ($p > f$) with all relevant rays for the graphical determination of an image position. As presented in figure 3.25. there are three relevant lens rays and there are their main characteristics:

- The ray that intersect the focal point F is refracted parallel to the optical axis,
- The ray that passes the lens center continues without refraction, and
- The ray that propagate parallel to the optical axis is refracted toward the focal point F.

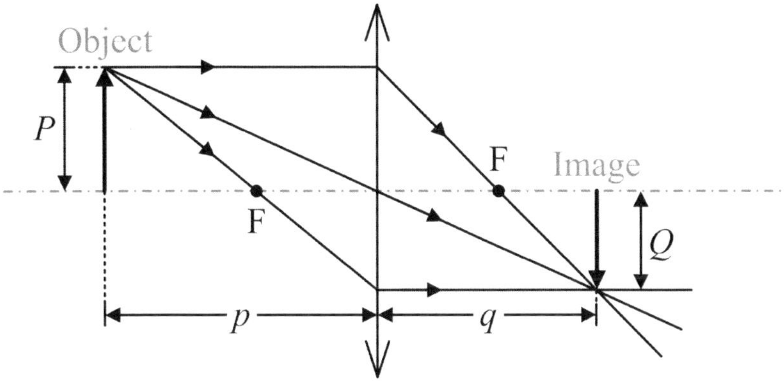

Figure 3.25. Thin convex lens image formation.

The image position is determined by the intersection of these rays and also by the equation (3.65). Similarly to the spherical mirror, the image magnification is given by:

$$M = \frac{Q}{P} = \frac{q}{p}. \qquad (3.68)$$

Thick Lens

If the lens thickness is comparable with other lens dimensions we cannot omit from the calculations the influence of the lens thickness. In figure 3.26. is presented the geometry of the thick lens with all relevant geometrical parameters. The lens thickness that will be taken into account is the distance between the lens vertexes V_1 and V_2, i.e. $V_1V_2 = t$. The lensmaker's equation for a thick lens in the air is given by:

$$\frac{1}{f} = (n-1)\left[\frac{1}{R_1} - \frac{1}{R_2} + \frac{(n-1)t}{nR_1R_2}\right]. \qquad (3.69)$$

As shown in figure 3.26. there are two principal planes where their intersections with the optical axis are marked with the points H_1 and H_2. These two planes have the property that a ray impinging the lens behaves in the same way as if we consider that the ray has no refraction prior to the arrival at the first principle plane and that the refraction occurs only between the front and rear principal planes. So, the refracted ray emanates from the rear principal plane. Therefore, the lens can be treated that all of the refraction appear at the principal planes. The importance of the principal planes is crucial in defining the optical system properties as the distance of the object and image from the front and rear principal planes must be taken into account. The distances between the lens vertexes and the principal planes are given by:

$$V_1H_1 = h_1 = -\frac{f(n-1)t}{nR_2}$$

$$V_2H_2 = h_2 = -\frac{f(n-1)t}{nR_1} \quad . \qquad (3.70)$$

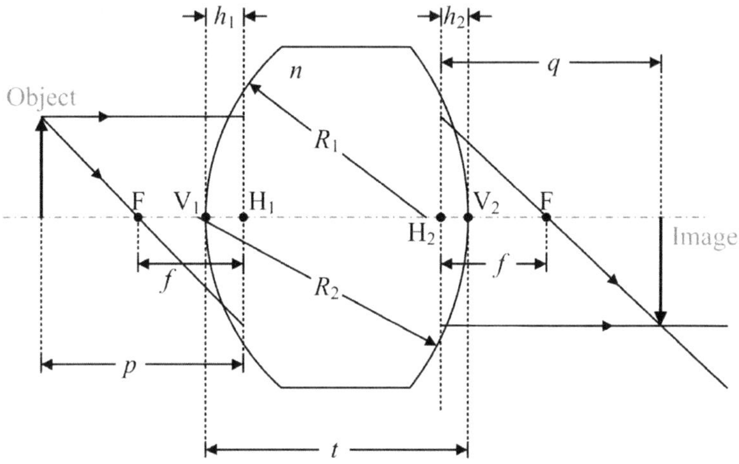

Figure 3.26. Thick convex lens.

MATRIX METHOD IN PARAXIAL OPTICS

When the optical system consists of several reflecting and/or refracting optical components the mathematical analysis becomes complex and time consuming. In the case of paraxial optical system a systematic approach in analyzing of such an optical system can be implemented which is based on the matrix method. Each ray transformation can be described by the corresponding matrix operator where the characterization of the overall optical system can be obtained by simple multiplication of individual matrices. For example, the optical system, presented in figure 3.27., consists of three optical lenses and one spherical mirror can be analyzed by implementing corresponding lens and mirror equations in order to finally find the image parameters (position and magnification). However, the matrix method has different approach. In order to find image parameters one can observe a ray that emanates from the object at the vertical position y_0 and angle θ_0 with respect to the optical axis. This ray passes through the several transformations between each characteristic point marked from 1 to 14 until the image has been formed.

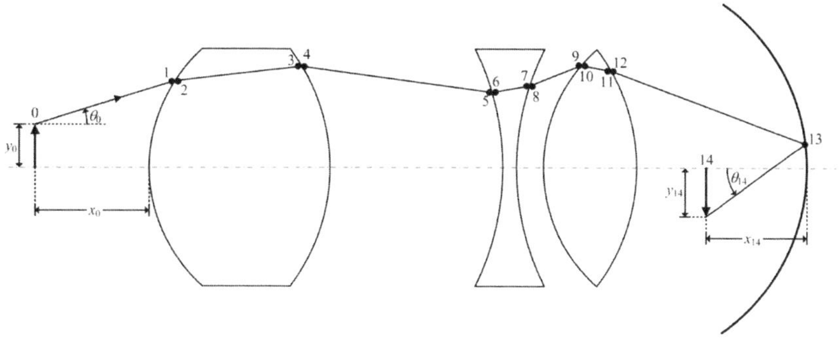

Figure 3.27. Optical system and matrix method.

From the point 0 to the point 1 we have simple ray translation, which is individually presented in figure 3.28. From the ray geometry one can write the following:

$$\theta_1 = \theta_0$$
$$y_1 = y_0 + d\tan\theta_0 \approx y_0 + d\theta_0, \tag{3.71}$$

where d is the ray translation distance and where in the paraxial approximation $\tan\theta_0 \approx \theta_0$ is valid.

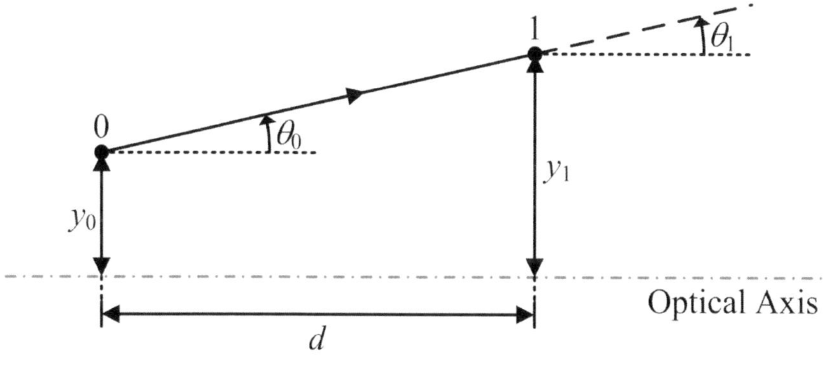

Figure 3.28. Ray translation.

Equation (3.71) can be rearranged in the following way:

$$y_1 = (1)y_0 + (d)\theta_0$$
$$\theta_1 = (0)y_0 + (1)\theta_0 \quad , \qquad (3.72)$$

or in the matrix notation:

$$\begin{bmatrix} y_1 \\ \theta_1 \end{bmatrix} = \begin{bmatrix} 1 & d \\ 0 & 1 \end{bmatrix} \begin{bmatrix} y_0 \\ \theta_0 \end{bmatrix} , \qquad (3.73)$$

where the translation matrix is given by:

$$\mathbf{M} = \begin{bmatrix} 1 & d \\ 0 & 1 \end{bmatrix} , \qquad (3.74)$$

Further, the ray is refracted between the points 1 and 2 on the curved surface as presented in figure 3.29. From the ray geometry one can write the following:

$$y_1 = y_0$$
$$\alpha_1 = \theta_1 + \varphi \approx \theta_1 + \frac{y_1}{R}$$
$$\alpha_0 = \theta_0 + \varphi \approx \theta_0 + \frac{y_0}{R} \quad , \qquad (3.75)$$

where R is the radius of curvature.

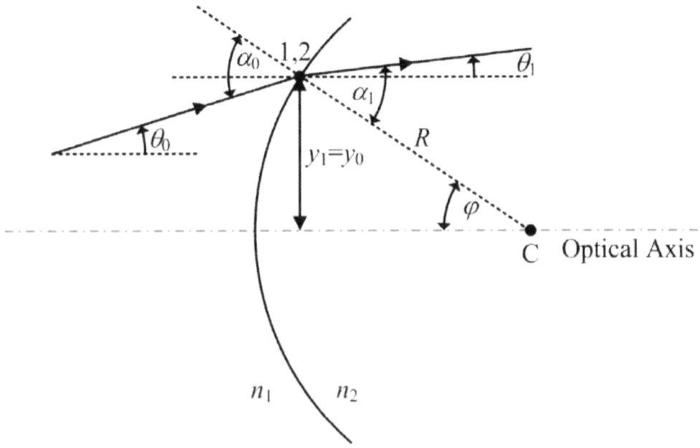

Figure 3.29. Ray refraction.

According to the Snell's law, we have:

$$n_2\alpha_1 = n_1\alpha_0 \, , \qquad (3.76)$$

where n_1 and n_2 are the corresponding indexes of refraction. From equations (3.75) and (3.76) we have:

$$\theta_1 = \frac{1}{R}\left(\frac{n_1}{n_2} - 1\right) y_0 + \frac{n_1}{n_2}\theta_0 . \quad (3.77)$$

Therefore, the appropriate linear equations are then:

$$y_1 = (1)y_0 + (0)\theta_0$$

$$\theta_1 = \left[\frac{1}{R}\left(\frac{n_1}{n_2} - 1\right)\right] y_0 + \left(\frac{n_1}{n_2}\right)\theta_0 , \quad (3.78)$$

or in the matrix notation:

$$\begin{bmatrix} y_1 \\ \theta_1 \end{bmatrix} = \begin{bmatrix} 1 & 0 \\ \frac{1}{R}\left(\frac{n_1}{n_2} - 1\right) & \frac{n_1}{n_2} \end{bmatrix} \begin{bmatrix} y_0 \\ \theta_0 \end{bmatrix} , \quad (3.78)$$

where the refraction matrix is given by:

$$\mathbf{M} = \begin{bmatrix} 1 & 0 \\ \frac{1}{R}\left(\frac{n_1}{n_2} - 1\right) & \frac{n_1}{n_2} \end{bmatrix} . \quad (3.79)$$

According to figure 3.27. from point 0 to point 13 we have only translation and refraction of the ray. However, at the point 7 the reflection from the spherical surface occurs as presented in figure 3.30.

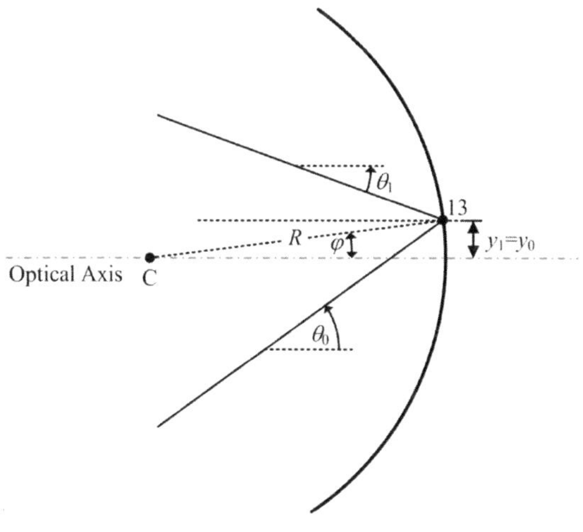

Figure 3.30. Ray reflection.

From the ray geometry and the law of reflection one can write the following:

$$y_1 = y_0$$
$$\theta_0 - \varphi = \theta_1 + \varphi \,, \tag{3.80}$$

where due to the radius sign convention the following is valid:

$$\varphi = -\frac{y_0}{R} \,. \tag{3.81}$$

By combining last two equations we have:

$$y_1 = (1)y_0 + (0)\theta_0$$
$$\theta_1 = \left(\frac{2}{R}\right)y_0 + (1)\theta_0 \,, \tag{3.82}$$

or in the matrix notation:

$$\begin{bmatrix} y_1 \\ \theta_1 \end{bmatrix} = \begin{bmatrix} 1 & 0 \\ \frac{2}{R} & 1 \end{bmatrix} \begin{bmatrix} y_0 \\ \theta_0 \end{bmatrix} , \tag{3.83}$$

where the reflection matrix is given by:

$$\mathbf{M} = \begin{bmatrix} 1 & 0 \\ \frac{2}{R} & 1 \end{bmatrix} . \tag{3.84}$$

If we return back to figure 3.27., we have the following ray transformations between the object and the image:

$$\begin{bmatrix} y_{14} \\ \theta_{14} \end{bmatrix} = \mathbf{M}_{14}\,\mathbf{M}_{13} \cdots \mathbf{M}_2 \mathbf{M}_1 \begin{bmatrix} y_0 \\ \theta_0 \end{bmatrix} , \tag{3.85}$$

where the ray parameters y_{14} and θ_{14} at the image position are obtained by multiplication of all particular matrix $\mathbf{M}_i$ starting from the first ray transformation. In general case, we can write the following imaging matrix equation:

$$\begin{bmatrix} y_\mathrm{i} \\ \theta_\mathrm{i} \end{bmatrix} = \begin{bmatrix} A & B \\ C & D \end{bmatrix} \begin{bmatrix} y_\mathrm{o} \\ \theta_\mathrm{o} \end{bmatrix} = \mathbf{M}_\mathrm{e} \begin{bmatrix} y_\mathrm{o} \\ \theta_\mathrm{o} \end{bmatrix} , \tag{3.86}$$

where $\mathbf{M}_\mathrm{e}$ is the equivalent matrix with the parameters A, B, C, and D, and y_i, θ_i and y_o, θ_o are the image and the object ray parameters, respectively. The equation (3.86) can be rearranged as:

$$y_i = Ay_o + B\theta_o$$
$$\theta_i = Cy_o + D\theta_o . \quad (3.87)$$

The imaging condition is fulfilled if it is satisfied $B = 0$ meaning that all the rays that emanate from one object points terminate also in a single image point. In this case the parameter A determines the optical system magnification. Table 3.3. describes the characteristics of an optical system based on the values of the parameters A, B, C, and D of the equivalent overall ray transformation matrix.

Table 3.3. A, B, C, and D parameter values and the optical system characteristics

Parameter	Characteristics
$A = 0$	All rays entering the input plane with the same angle pass through the same point at the output plane. The output plane represents the focal point of an optical system.
$B = 0$	All rays leaving one point at the input plane will end in a single point at the output plane. The input plane represents an object and the output plane represents an image of an optical system.
$C = 0$	All rays entering the input plane with the same angle emerge at the output plane with the same, but in general case different angle then at the input plane. Such an optical system is called an afocal or telescopic system.
$D = 0$	All rays entering the input plane from the same point emerge at the output plane with the same angle. The input plane represents the focal point of an optical system.

The summary of ray transformation matrixes for typical optical components has been presented in table 3.4.

Table 3.4. Summary of ray transformation matrixes for typical optical components

Translation matrix	$\mathbf{M} = \begin{bmatrix} 1 & d \\ 0 & 1 \end{bmatrix}$	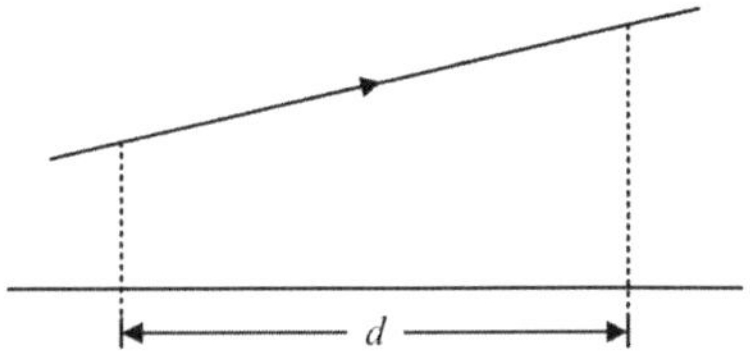
Spherical surface refraction matrix	$\mathbf{M} = \begin{bmatrix} 1 & 0 \\ \frac{1}{R}\left(\frac{n_1}{n_2} - 1\right) & \frac{n_1}{n_2} \end{bmatrix}$	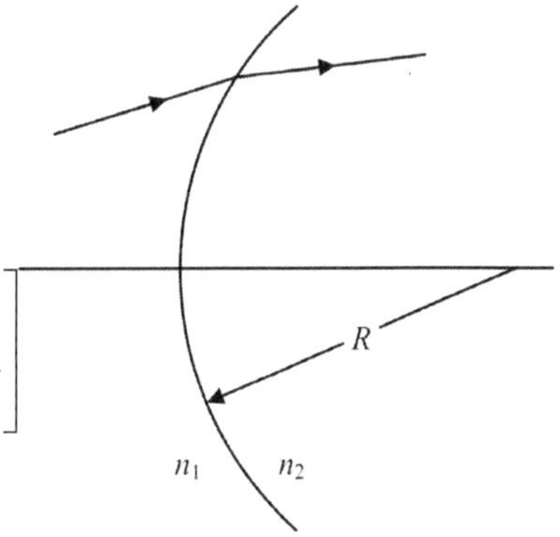
Plane surface refraction matrix	$\mathbf{M} = \begin{bmatrix} 1 & 0 \\ 0 & \frac{n_1}{n_2} \end{bmatrix}$	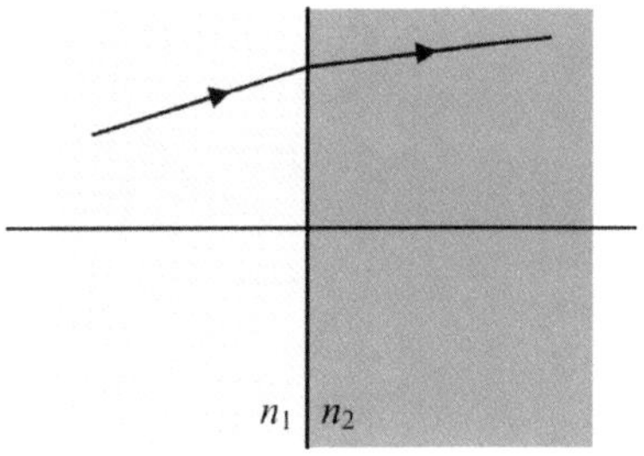
Thin lens matrix	$\mathbf{M} = \begin{bmatrix} 1 & 0 \\ -\frac{1}{f} & 1 \end{bmatrix}$	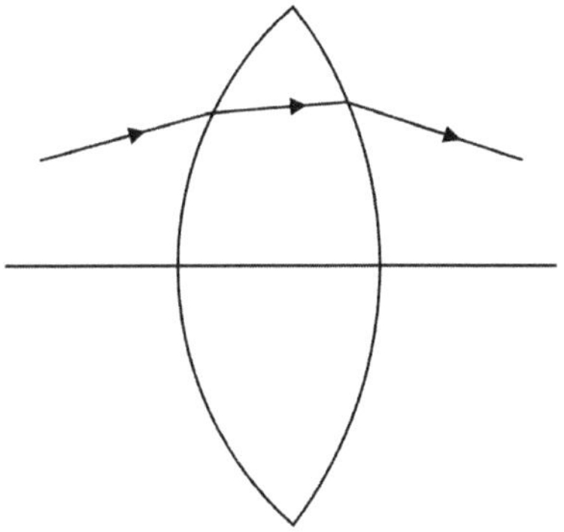

Spherical mirror matrix	$\mathbf{M} = \begin{bmatrix} 1 & 0 \\ \frac{2}{R} & 1 \end{bmatrix}$	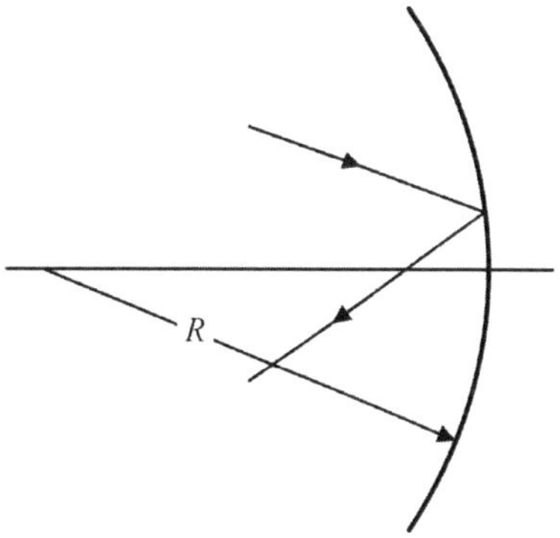

OPTICAL INSTRUMENTS

Geometrical optics represents a powerful tool in analyzing and designing of an optical instrumentation. Here it will be analyzed different types of optical instruments that are typically used for linear and angular magnification as well as in optical spectrum analysis.

Prism Spectrometer

If a prism is used as a dispersive element a simple optical spectrometer can be built as it is presented in figure 3.31. Bearing in mind that ray deviation angle depends on the prism index of refraction which further is dependent on the wavelength one can characterize the source spectrum by simple measuring the captured flux versus the deviation angle. In order to build such a spectrometer the light, emanating from the source to be characterized, must be collimated by the collimator and after the double refraction at the prism viewed by the telescope focused for infinity. The prism is typically mounted on a rotatable platform in order to perform angular scanning.

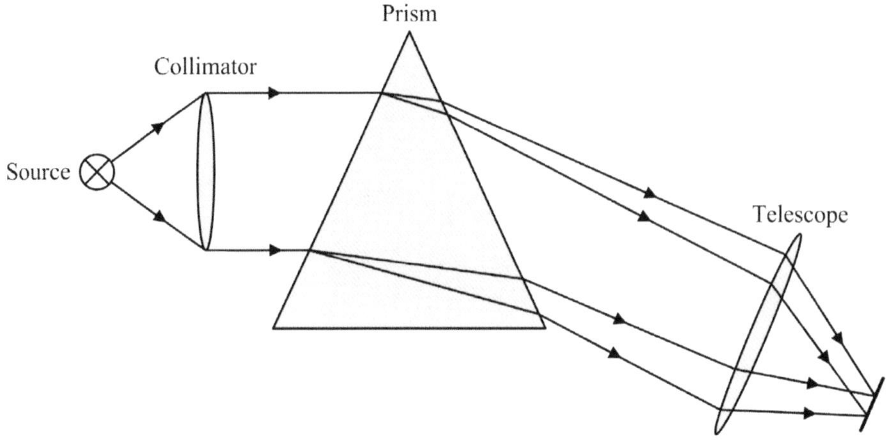

Figure 3.31. Prism spectrometer.

Such a presented optical spectrometer has the finite resolution which is determined by the spectrometer geometry and the prism material characteristics. The resolving power of a spectrometer represents an important parameter that can determine the spectrometer choice. In order to determine the prism spectrometer resolution, i.e. the minimal difference between two wavelengths that can be resolved at the spectrometer output, we will consider an optical setup which is depicted in figure 3.32.

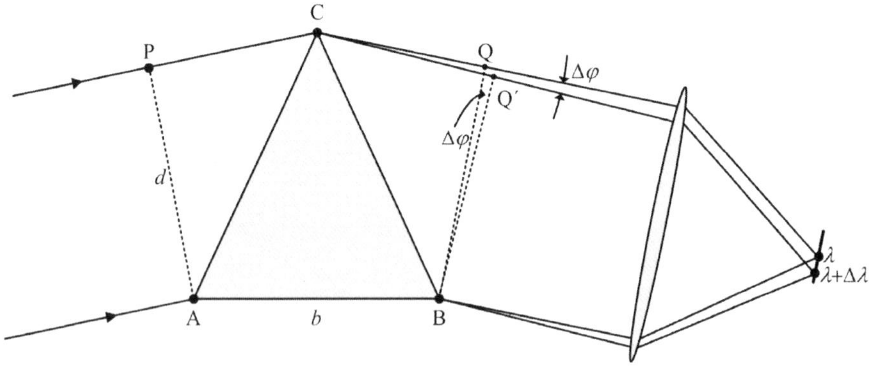

Figure 3.32. Prism spectrometer resolution.

Based on the Fermat's principle and taking into consideration that lines AP and BQ belong to the corresponding eqiphase planes we have:

$$PC + CQ = n(\lambda)\cdot AB = n(\lambda)b, \qquad (3.88)$$

where $n(\lambda)$ is the prism index of refraction at the given wavelength λ and b prism base length. The similar situation we have for the wavelength $\lambda+\Delta\lambda$, where according to the Fermat's principle we also have:

$$\mathrm{PC} + \mathrm{Q} \quad + \mathrm{QQ}' = n(\lambda + \Delta\lambda)b \,. \qquad (3.89)$$

From the last two equations we have:

$$\mathrm{QQ}' = [n(\lambda + \Delta\lambda) - n(\lambda)]b \approx b\left(\frac{\mathrm{d}n}{\mathrm{d}\lambda}\right)\Delta\lambda \,. \qquad (3.90)$$

Geometry from figure 3.32. us:

$$\mathrm{QQ}' \approx d\Delta\varphi \,, \qquad (3.91)$$

where $d = \mathrm{AP}$ is the beam width and $\Delta\varphi$ is the angular difference. Based on the Rayleigh's criterion for the diffraction limited imaging resolution, we have:

$$\Delta\varphi \approx \frac{\lambda}{d} \,. \qquad (3.92)$$

The combination of the equations (3.90), (3.91) and (3.92) gives:

$$\Delta\lambda_{\mathrm{min}} \approx \frac{\lambda}{b\left(\frac{\mathrm{d}n}{\mathrm{d}\lambda}\right)} \,, \qquad (3.93)$$

where $\Delta\lambda_{\mathrm{min}}$ represents the minimum wavelength separation. By definition the resolving power is given by:

$$\Re = \frac{\lambda}{\Delta\lambda_{\mathrm{min}}} = b\left(\frac{\mathrm{d}n}{\mathrm{d}\lambda}\right) \,. \qquad (3.93)$$

Magnifying Glasses and Eyepieces

A magnifying glass in its simplest form is typically a convex lens which purpose is to magnify an image of a small object. Sometimes a doublet or triplet can be used in order to perform higher quality imaging. Magnifying glasses typically have low magnifying power somewhere in the range from 2 to 6. Simple magnifying glasses with higher magnifications produce images with poor quality due to optical aberrations, especially due to the spherical aberration. Figure 3.33. represents the working principle of a simple magnifying glass.

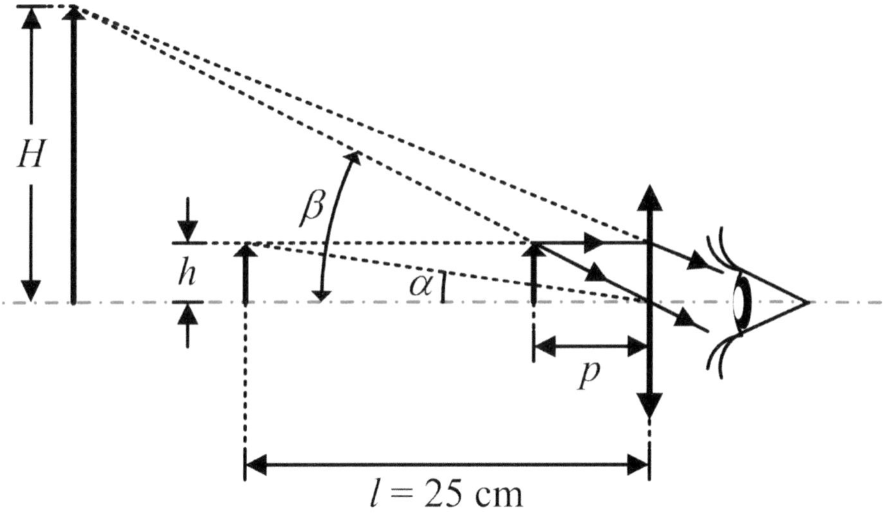

Figure 3.33. Magnifying glass.

If we have a small object with the height h it is assumed that it should be positioned at the distance l = 25 cm from the unaided eye, i.e. at the near point of the normal eye (nearest point of distinct vision). At this point the eye sees the object with the angle α. In order to magnify the object image one can position object closer to the eye with the lens in between. In this position, lens forms a virtual image with the viewing angle β. The angular magnification is defined as the ratio of these to angles:

$$M = \frac{\beta}{\alpha} = \frac{\frac{h}{p}}{\frac{h}{l}} = \frac{l}{p} \quad (3.94)$$

If the image is viewed at the infinity we have $p = f$, where f is the lens focal length and thus the magnification is given by:

$$M = \frac{l}{f} \quad (3.95)$$

The other extreme situation is when the virtual image is also at the position of the nearest point of the normal eye i.e. $q = -l$, where q is the virtual image position. According to the lens equation in this situation the object must be at the position:

$$p = \frac{f}{l+f}, \tag{3.96}$$

And the corresponding magnification is given by:

$$M = 1 + \frac{l}{f}. \tag{3.97}$$

The actual magnification depends on the particular viewer which will position the lens with the respect of the object and eye in order to obtain appropriate virtual image.

In order to achieve higher angular magnifications, typically greater then 10, a doublet or triplet must be used which are called oculars or eyepieces. To provide high quality images the ocular (eyepiece) is to some extent corrected for aberrations especially to reduce transverse chromatic aberration. The chromatic aberration is typically courted with the help of a doublet. The effective focal length of the doublet is given by:

$$\frac{1}{f} = \frac{1}{f_1} + \frac{1}{f_2} - \frac{d}{f_1 f_2}, \tag{3.98}$$

where f_1 and f_2 are the focal lengths of the lenses and d is its mutual distance. According to the lensmaker's equation, the individual focal lengths are given by:

$$\frac{1}{f_1} = (n-1)\left(\frac{1}{R_1} - \frac{1}{R_2}\right) =$$

$$\frac{1}{f_2} = (n-1)\left(\frac{1}{R_1} - \frac{1}{R_2}\right) = (n-1)C_2, \tag{3.99}$$

where R_{ij} (i = 1,2) are the corresponding lens curvatures and C_1 and C_2 the corresponding parameters. Incorporating equation (3.99) into (3.98) we have:

$$\frac{1}{f} = (n-1)C_1 + (n-1)C_2 - d(n-1)^2 C_1 C_2 . \tag{3.100}$$

The correction of the transverse chromatic aberration is achieved if the overall focal length is independent on the change of the index of refraction, i.e. if it is satisfied:

$$\frac{\mathrm{d}\left(\frac{1}{f}\right)}{\mathrm{d}n}=0, \qquad (3.101)$$

where from equations (3.100) and (3.101) we have:

$$C_1+C_2-2d(n-1)C_1C_2=0, \qquad (3.102)$$

which is fulfilled if it is satisfied:

$$d=\frac{C_1+C_2}{2(n-1)C_1C_2}, \qquad (3.103)$$

Or simpler when it is fulfilled:

$$d=\frac{1}{2}(f_1+f_2). \qquad (3.104)$$

This last condition if fulfilled for the Huygens and Ramsden eyepiece.

Beside the transverse chromatic aberration there is also the spherical aberration that is present especially at the high magnification glasses typically with the magnification larger than 10. A Coddington magnifier is a high magnifying glass which consists of a single very thick lens with a central deep groove which limits the rays to those close to the axis (paraxial rays), which further suppress the spherical aberration. Therefore, greater magnification than a conventional magnifying glass can be achieved, typically 10 up to 20. The drawback is that the diaphragm groove reduces the field of view of the magnifier. A typical cross section of the Coddington magnifier is presented in figure 3.34.

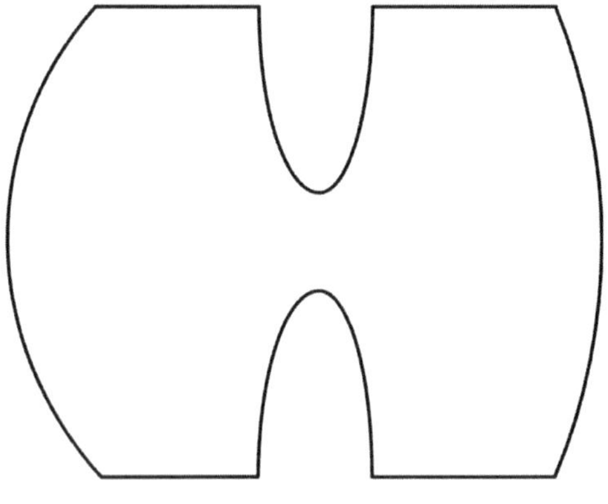

Figure 3.34. Coddington magnifier.

Microscopes

Large magnification of a small object can be accomplished by the compound microscope as it is presented in figure 3.35. According to the optical setup the object is placed at the distance larger than the focal length of the objective and a real intermediate image is placed within the microscope tube. The position of the intermediate image is close to the eyepiece focal point. In this case the eye positioned near to the eyepiece sees a virtual image that is magnified.

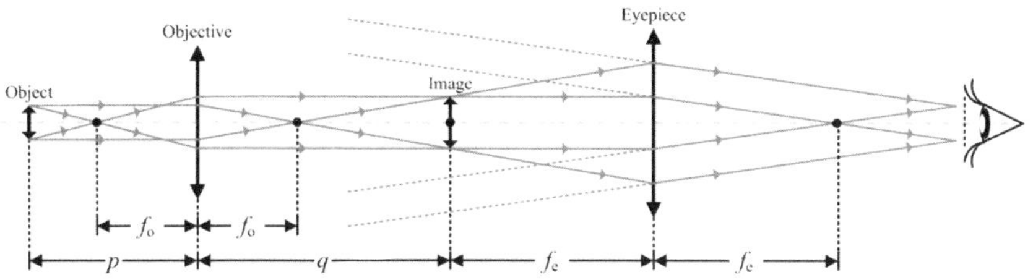

Figure 3.35. Microscope image formation.

When the final image is seen by the eye the overall magnification of the microscope can be determined as in the case of a simple magnifying glass. Therefore, the angular magnification of the microscope is given by:

$$M = \frac{l}{f_{\mathrm{eff}}}, \qquad (3.105)$$

where $l = 25$ cm and f_{eff} is the effective focal length of the objective and eyepiece given by:

$$\frac{1}{f_{\mathrm{eff}}} = \frac{1}{f_{\mathrm{o}}} + \frac{1}{f_{\mathrm{e}}} - \frac{d}{f_{\mathrm{o}} f_{\mathrm{e}}}, \qquad (3.106)$$

where f_{o} is the objective focal length, f_{e} is the eyepiece focal length and d is the distance between the objective and eyepiece. Substituting equation (3.1067) into equation (3.105) we have:

$$M = \frac{l(f_{\mathrm{o}} + f_{\mathrm{e}} - d)}{f_{\mathrm{o}} f_{\mathrm{e}}}. \qquad (3.107)$$

Taking into consideration the thin lens equation for the objective lens we have:

$$\frac{q}{p} = \frac{d - f_o - f_e}{f_{oe}} \quad , \qquad (3.108)$$

where $q = d\text{-}f_e$ is evident form figure 3.35. The combination of the last two equations gives:

$$M = -\frac{l}{f_e}\frac{q}{p} \quad , \qquad (3.109)$$

which indicates that the microscope magnification is a product of individual magnifications of the objective and the eyepiece.

Telescopes

The basic function of a telescope is to magnify the apparent size of a distant object. This is accomplished by enlarging the angle at which the observer sees the image of a distant object. The telescope magnification or telescope power is simply the ratio of the angle subtended by the image to the angle subtended by the object. Typically, a telescope has both its object and image located at infinity, so it represents an afocal optical instrument, since it has no focal length. Below it will be presented the basic relationships for telescopes and afocals.

As presented in figure 3.36. in principle there are three major types of telescopes: astronomical (inverting or Keplerian), terrestrial (erecting), and Galilean. An astronomical telescope is composed of two positive lenses spaced in a way that their mutual distance is equal to the sum of both focal lengths, as presented in figure 3.36 (a). The objective lens forms an inverted image at its focal point where the eyepiece then reimages the object at infinity where it may be comfortably observed by a relaxed eye. As the intermediate image is inverted, and the eyepiece doesn't reinvert the image, the eye sees the inverted image.

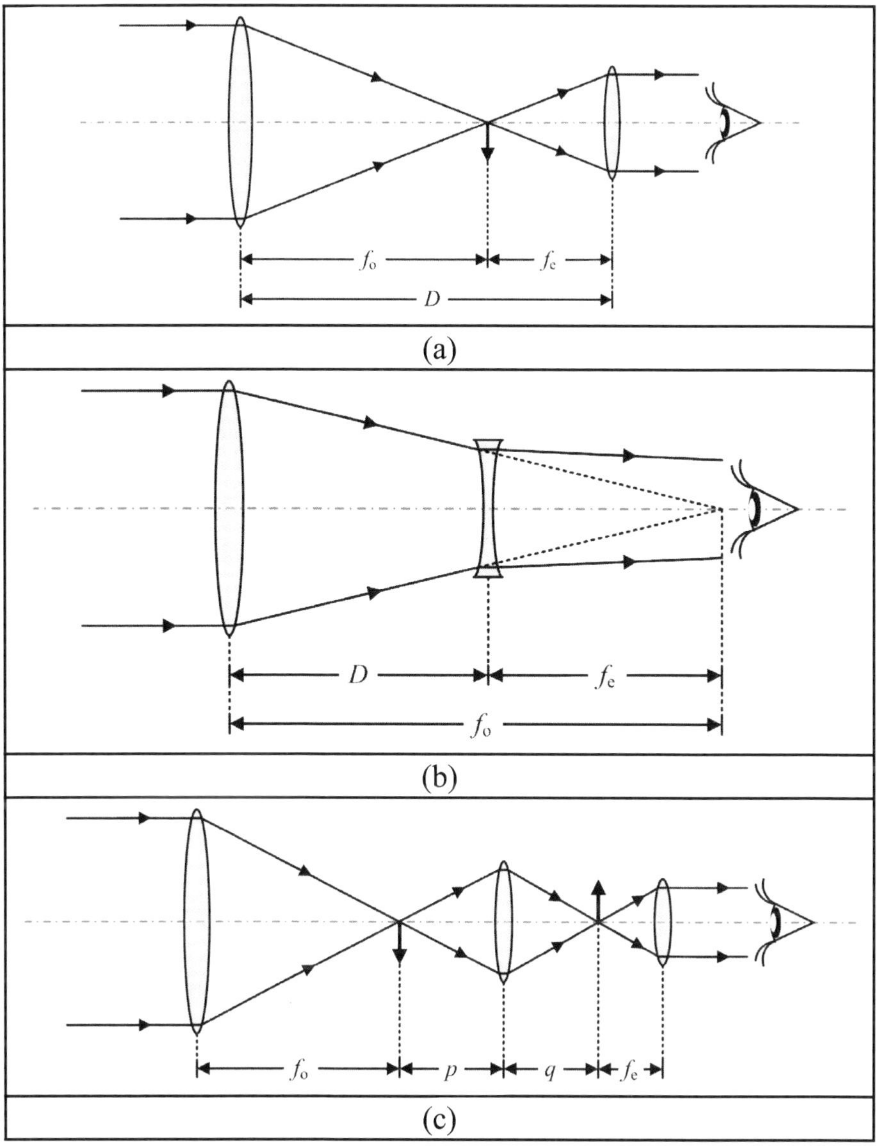

Figure 3.36. Three basic types of telescope: (a) astronomical telescope, (b) Galilean telescope, and (c) lens erecting telescope.

As presented in figure 3.36 (b) in a Galilean telescope the positive eyepiece lens is replaced by a negative where the spacing between the lenses is equal to the objective and eyepiece lens focal lengths difference. The intermediate image isn't formed, so the object for the eyepiece is a virtual object. Therefore, no inversion occurs, and the final image as seen by the eye is erect and unreversed.

Considering the telescope components to be thin lenses one can derive important relationships that can be applied to the telescopes and afocal optical systems. Bearing in mind that the focal length of a negative lens is negative the distance between the objective and eyepiece can be represented as the sum of both focal lengths, i.e. as:

$$D = f_o + f_e \, . \qquad (3.110)$$

The magnification of the telescope is the ratio between the angle subtended by the image, and the angle subtended by the object. The intermediate image size, formed by the objective, will be given as:

$$h = \alpha_o f_o \, , \qquad (3.111)$$

where α_o is the angle subtended by the object. The angle subtended by this image from the eyepiece will be:

$$\alpha_e = -\frac{h}{f_e} \, . \qquad (3.112)$$

From the combination of equations (3.111) and (3.112) we get the telescope magnification as:

$$M = \frac{\alpha_e}{\alpha_o} = -\frac{f_o}{f_e} \, . \qquad (3.113.)$$

REFERENCES

1. Vladimir Protopopov. *Practical Opto-Electronics*. Cham: Springer International Publishing Switzerland, 2014.
2. Oleg Marchenko, Sergei Kazantsev, and Laurentius Windholz. *Demonstrational Optics - Part 2: Coherent and Statistical Optics*. New York: Springer Science+Business Media, LLC, 2007.
3. Keigo Iizuka. *Engineering Optics (3rd Edition)*. New York: Springer Science+Business Media, LLC, 2008.
4. Antonio Romano. *Geometric Optics: Theory and Design of Astronomical Optical Systems Using Mathematica* ®. New York:

Birkhaüser Boston, a part of Springer Science+Business Media, LLC, 2010.

5. Eustace L. Dereniak and Teresa D. Dereniak. *Geometrical and Trigonometric Optics*. Cambridge: Cambridge University Press, 2008.
6. Germain Chartier. *Introduction to Optics*. New York: Springer Science+Business Media, Inc., 2005.
7. Daniel Malacara and Zacarias Malacara. *Handbook of Optical Design (2nd Edition)*. New York: Marcel Dekker, Inc., 2004.
8. Frank L Pedrotti, Leno M Pedrotti, and Leno S Pedrotti. *Introduction to Optics (3rd Edition)*. Harlow: Pearson Education Limited, 2014.
9. Kjell J. Gåsvik. *Optical Metrology (3rd Edition)*. Chichester: John Wiley & Sons Ltd, 2002.
10. Warren J. Smith. *Modern Optical Engineering: The Design of Optical Systems (3rd Edition)*. New York: The McGraw-Hill Companies, Inc., 2000.
11. Mortimer Abramowitz. *Microscope Basics and Beyond (Revised Edition 2003)*. Melville: Olympus America Inc., 2003.
12. Warren J. Smith. *Modern Optical Engineering: The Design of Optical Systems (4th Edition)*. New York: The McGraw-Hill Companies, Inc., 2008.
13. Max J. Riedl. *Optical Design: Applying the Fundamentals*. Bellingham: Society of Photo-Optical Instrumentation Engineers, 2009.
14. Bruce H. Walker. *Optical Engineering Fundamentals (2nd Edition)*. Bellingham: The Society of Photo-Optical Instrumentation Engineers, 2008.
15. Robert E. Fischer, Biljana Tadic-Galeb, and Paul R. Yoder. *Optical System Design*. New York: The McGraw-Hill Companies, Inc., 2008.
16. B. K. Johnson. *Optics and Optical Instruments: An Introduction with Special Reference to Practical Applications*. New York: Dover Publications, Inc., 1960.
17. Dieter Meschede. *Optics, Light and Lasers: The Practical Approach*

to Modern Aspects of Photonics and Laser Physics (2nd Revised and Enlarged Edition). Weinheim: WILEY-VCH Verlag GmbH & Co. KGaA, 2007.

18. K. K. Sharma. *Optics: Principles and Applications*. Amsterdam: Elsevier Inc., 2006.
19. Hartmann Römer. *Theoretical Optics: An Introduction*. Weinheim: WILEY-VCH Verlag GmbH & Co. KGaA., 2005.

4

CHAPTER

COHERENCE AND INTERFERENCE OF LIGHT

TWO BEAM INTERFERENCE

A light ray, which was used in previous chapter in describing the geometrical optics phenomena, fails trying to explain optical phenomena like diffraction and interference. Therefore, instead of a light ray we will take into consideration in this chapter the wave nature of electromagnetic radiation. For the purpose of simplicity we will consider only a linearly polarized plane wave propagating in a vacuum in the *z* direction. In this case the electric at any point along the direction of propagation and any time moment can be represented by a harmonic function of distance and time as:

$$E(z,t) = E_0 \cos(\omega t - k\ ), \qquad (4.1)$$

where E_0 is the electrical field amplitude, k is the wave vector and ω is the angular frequency. Sometime it is more convenient to represent an electrical field of a plane wave by using Euler representation of complex number as:

$$E(z,t) = \mathrm{R}\ \{E_0 \exp[j(\omega t - k\ )]\}, \qquad (4.2)$$

where $\mathrm{Re}\{\bullet\}$ represents the real part of a complex number. If the mathematical operations on electrical field are linear then, without losing the meaning of the analysis, we can represent the electrical field as a complex number in the following form:

$$E(z,t) = E_0 \exp[j(\omega t - kz)], \tag{4.3}$$

and take a real part of the final result at the end of the calculation. Equation (4.3) can be reorganized as:

$$E(z,t) = E_0 \exp(-jkz)\exp(j\omega t) = A\exp(j\omega t), \tag{4.4}$$

where A is the complex amplitude given by:

$$A = E_0 \exp(-jkz). \tag{4.5}$$

Due to the extremely high frequencies of optical waves (in the visible range we typically have $\nu \approx 6\times10^{14}$ Hz for $\lambda = 500$ nm) direct measurement of the electric field is not possible. It is only possible to measure directly the radiation parameters which are related to the optical wave's energetic content. Therefore, one of the most important quantities is the irradiance, which represents the time average of the amount of energy which, in unit time, crosses a unit area normal to the direction of the energy flow. The irradiance is thus proportional to the time average of the square of the electric field, i.e.:

$$I = \langle E^2(t) \rangle = \lim_{T\to\infty} \frac{1}{T}\int_0^T E^2(t)\,dt. \tag{4.6}$$

The combination of equation (4.1), and (4.6)

$$I = \lim_{T\to\infty} \frac{1}{T}\int_0^T E_0^2 \cos^2(\omega t - kz)\,dt = \frac{E_0^2}{2} = \frac{|A|^2}{2}. \tag{4.7}$$

Ignoring the scale factor ½, the important conclusion that can be extract is that the irradiance is directly proportional to the electrical field amplitude squared.

In the case of two monochromatic waves, which propagate in the same direction and polarized in the same plane, a superposition of the waves occurs at a certain point in space, where the total electric field at this point is given by:

$$E = E_1 + E_2, \tag{4.8}$$

where E_1 and E_2 are the electric fields of two waves. In the case where both waves have the same frequency the irradiance at this point is:

$$I = |A_1 + A_2|^2, \qquad (4.9)$$

where $A_1 = E_0 \exp(-jkz_1) = E_0 \exp(-j\varphi_1)$ and $A_2 = E_0 \exp(-jkz_2) = E_0 \exp(-j\varphi_2)$ are the complex amplitudes of the two waves. In this case equation (4.9) becomes:

$$I = |A_1|^2 + |A_2|^2 + |A_1 A_2^*| + |A_1^* A_2|$$

$$I = I_1 + I_2 + 2\sqrt{I_1 I_2}\cos\Delta\varphi, \qquad (4.10)$$

where I_1 and I_2 are the intensities of the waves and $\Delta\varphi = \varphi_1 - \varphi_2$ is the phase difference between them. Depending on the phase difference, i.e. on the optical path difference between the waves, the irradiance at the point of observation can take any value between the maximum and minimum values that are given by:

$$I_{min} = I_1 + I_2 - 2\sqrt{I_1 I_2}$$

$$I_{max} = I_1 + I_2 + 2\sqrt{I_1 I_2}. \qquad (4.11)$$

An appropriate measure of the interference phenomenon contrast is the visibility of the interference fringes, which is defined as:

$$V = \frac{I_{max} - I_{min}}{I_{max} + I_{min}} = \frac{2\sqrt{I_1 I_2}}{I_1 + I_2}, \qquad (4.12)$$

which can take the any value between 0 and 1 ($0 \le V \le 1$).

If we take into consideration a light from a thermal source, even if it consists of a single spectral line, this light source is not strictly monochromatic. The reason for this is the random rapid fluctuations of the electric field amplitude and phase. So, if we observe two such independent light sources, which fluctuations are uncorrelated, interference effects cannot be observed. At this stage, we will consider that in order to obtain interference phenomena we must obtain two light beams from the same light source. Therefore, we shall consider that the observed interference phenomena between two beams can be classified according to the method used to obtain these beams.

Wavefront Division Method

In order to obtain two beams from a single light source one can use two portions of the original wavefront. Then, both of the beams are superimposed in order to obtain interference pattern. This method is known as wavefront division. A simple optical system based on the Fresnel's mirrors arrangement is presented in figure 4.1. The light from a point like source S has been used to illuminate two mirrors M_1 and M_2 which are mutually inclined for a very small angle α (α << 1). The reflected light is directed toward the screen where the beams are overlapped and thus the interference fringes occur on the screen. Equivalently we can consider that the interference take place between the light from the two virtual sources S_1 and S_2, i.e. the images of the source S in the mirrors M_1 and M_2. According to the presented mirrors and source geometry we have:

$$S_1S_2 \approx 2d\alpha \quad , \qquad (4.13)$$

where S_1S_2 is the distance between the virtual sources and d is the distance between the mirrors and the virtual sources. The optical path difference between the beams at the screen point positioned at the distance x from the optical is, to a first approximation given by:

$$OPD \approx \frac{2d\acute{a}}{d+D} \quad , \qquad (4.14)$$

where D is the distance between the mirrors and the screen. As the successive maxima or minima in the interference pattern correspond to a change in the optical path difference equals to the light source wavelength λ, the distance between the fringes is given by:

$$\Delta x = \frac{\lambda(d+D)}{2d\acute{a}} \quad . \qquad (4.15)$$

There are also other arrangements such as Young's double slit interferometer, presented in figure 4.2., which uses the diffracted beams from a pair of slits S_1 and S_2, and Lloyd's mirror, presented in figure 4.3., which uses one beam from the source S and another from a mirror illuminated at near-grazing incidence.

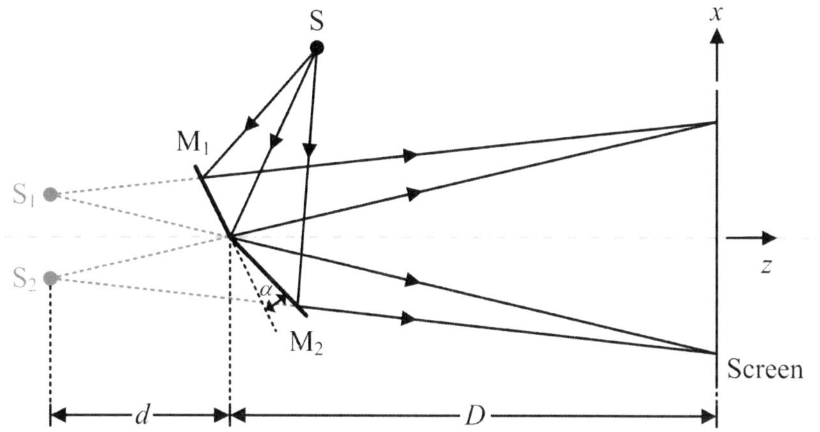

Figure 4.1. Optical interference produced by wavefront division (Fresnel's mirrors).

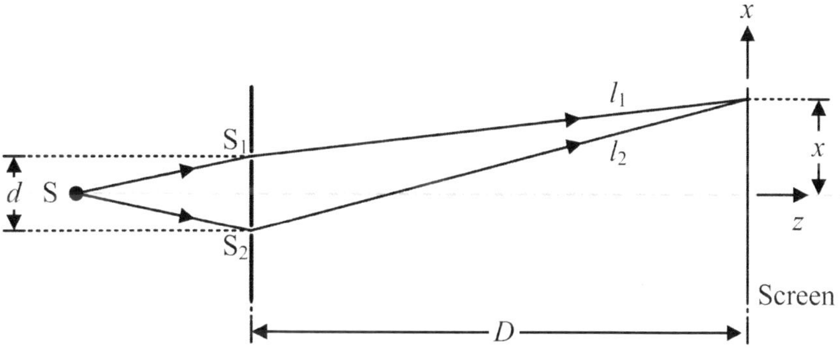

Figure 4.2. Young's double slit interferometer.

In the case of the Young's double slit interferometer, the optical path difference between the beams that impinge the screen at the point which is positioned at the distance x from the optical axis (axis of symmetry) is given by:

$$OPD \approx l_1 - l_2, \qquad (4.16)$$

where l_1 and l_2 are the beam lengths from the slits S_1 and S_2, respectively. These distances are given with:

$$l_1 = \sqrt{D^2 + \left(x - \frac{d}{2}\right)^2}$$

$$l_2 = \sqrt{D^2 + \left(x + \frac{d}{2}\right)^2}, \qquad (4.17)$$

where d is the distance between the slits and D is the distance between the slits and the screen. Equations (4.16) and (4.17) in the case when $D >> d, x$ give:

$$OPD \approx \sqrt{D^2 + \left(x - \frac{d}{2}\right)^2} - \sqrt{D^2 + \left(x + \frac{d}{2}\right)^2}$$

$$OPD \approx -\frac{xd}{D} \qquad (4.18)$$

As the successive maxima or minima in the interference pattern correspond to a change in the optical path difference equals to the light source wavelength λ, the distance between the fringes is given by:

$$\Delta x = \frac{D\lambda}{d}. \qquad (4.19)$$

In the case of the Lloyd's interferometer, the optical path difference between the beams that impinge the screen at the point which is positioned at the distance x from the optical axis (axis of symmetry) is given by:

$$OPD \approx l_1 - l_2, \qquad (4.20)$$

where l_1 is the beam length from the source S to the point at the screen located at the position x from the optical axis, and l_2 is the reflected beam length, which is equal to the length from the virtual source S′ to the observation point at the screen. These distances are given with:

$$l_1 = \sqrt{D^2 + (x - d)^2}$$

$$l_2 = \sqrt{D^2 + (x + d)^2}, \qquad (4.21)$$

where d is the distance between the source S and the horizontal mirror and D is the distance between the source and the screen. Equations (4.20) and (4.21) in the case when $D >> d, x$ give:

$$OPD \approx \sqrt{D^2 + (x-d)^2} - \sqrt{D^2 + (x+d)^2}$$
$$OPD \approx -\frac{2xd}{D} \quad . \quad (4.22)$$

As the successive maxima or minima in the interference pattern correspond to a change in the optical path difference equals to the light source wavelength λ, the distance between the fringes is given by:

$$\Delta x = \frac{D\lambda}{2d} \quad . \quad (4.23)$$

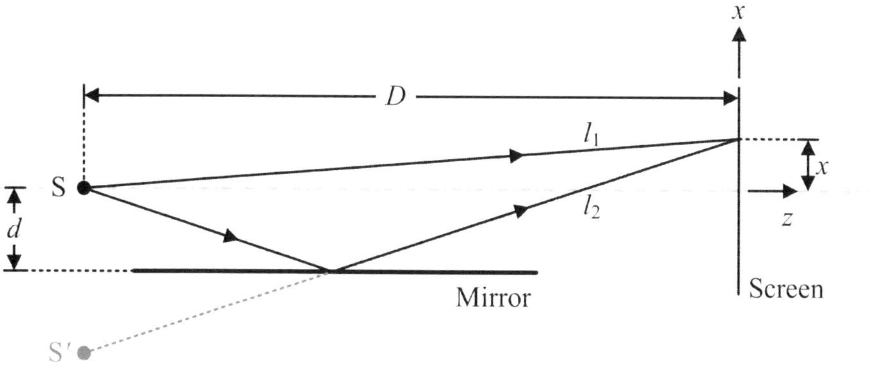

Figure 4.3. Lloyd's mirror.

Amplitude Division Method

The other method where two beams can be obtained from a single source is by division of the amplitude over the same section of the wavefront. In order to perform amplitude division of the wavefront there is one way where we can use a surface that partially reflects the incident light and partially transmits it. If a transparent plane-parallel plate is illuminated by the monochromatic point source S, as presented in figure 4.4., then at a given point at the observation plane two beams, reflected from the upper and lower surface of the plane-parallel plate, arrive approximately with the same amplitude. Therefore, from the setup symmetry, the observed interference fringes have circular shape. In the case when the observation plane is positioned at infinity the interference fringes can be obtained at the back focal plane of the lens, as shown in figure 4.5. The optical path difference between the beams is given by:

$$OPD \approx n(AB + BC) - AD, \qquad (4.24)$$

where n is the plan-parallel plate index of refraction, and lengths AB, BC, and AD are presented in figure 4.5.

The geometry from figure 4.5. also gives the following relations:

$$AB = BC = \frac{d}{\cos\theta_2}$$

$$AD = AC\ \sin\theta_1 = 2d\tan\theta_2\sin\theta_1, \qquad (4.25)$$

where d is the plate thickness. Taking into consideration Snell's law we have $\sin\theta_1 = n\sin\theta_2$ we have:

$$OPD \approx 2nd\ \cos\theta_2. \qquad (4.26)$$

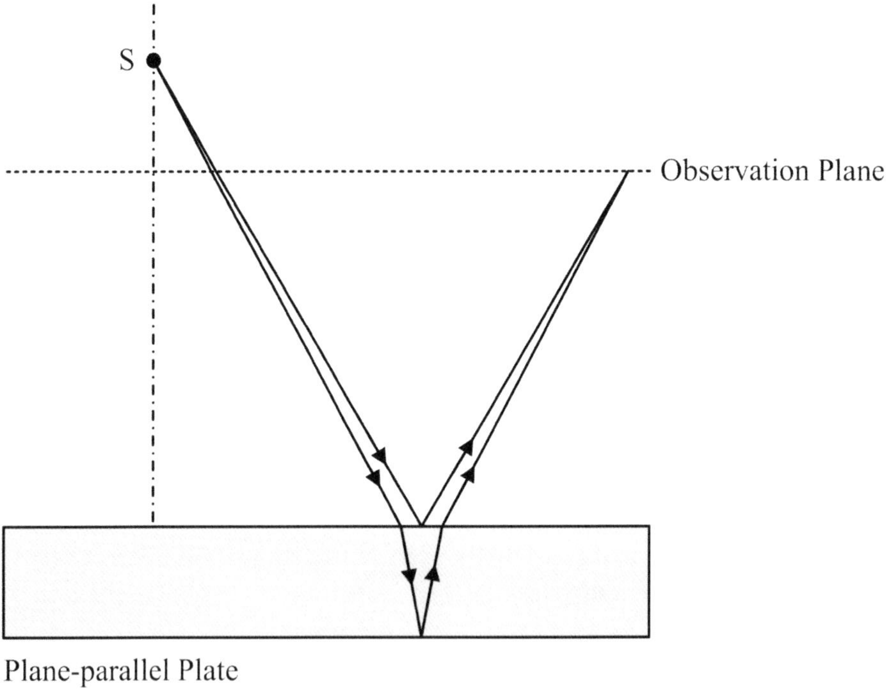

Figure 4.4. Interference fringes formation by the reflection in a plane-parallel plate.

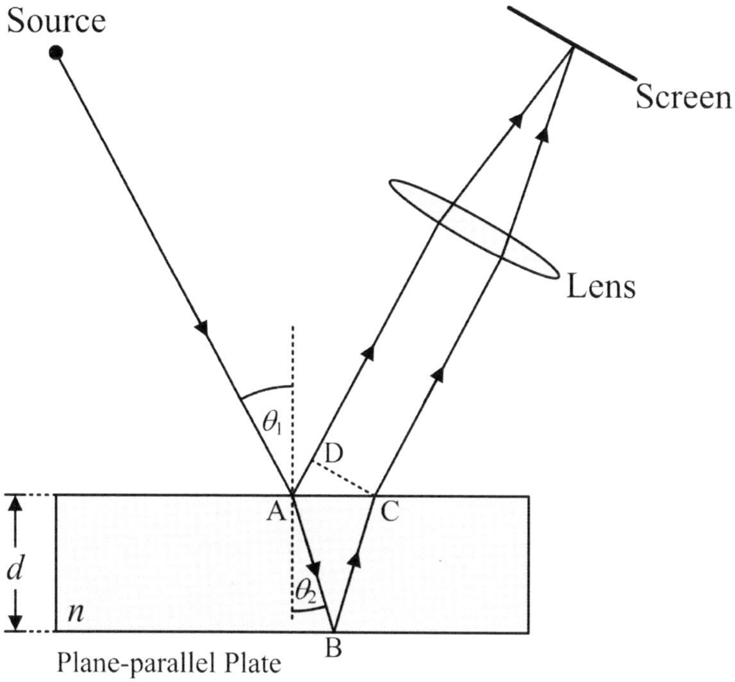

Figure 4.5. Interference fringes formation of equal inclination by reflection in a plane-parallel plate.

However, we also have to take into account a phase shift of π introduced by reflection at one of the plate surfaces. Therefore, the effective optical path difference between the interfering wavefronts is:

$$OPD \approx 2nd \cos\theta_2 \pm \frac{\lambda}{2}. \quad (4.27)$$

The conditions for the bright and dark fringes are given:

$$OPD \approx 2nd \cos\theta_2 \pm \frac{\lambda}{2} = k\lambda$$

$$OPD \approx 2nd \cos\theta_2 \pm \frac{\lambda}{2} = (2k+1)\frac{\lambda}{2}, \quad (4.28)$$

respectively, where k is the integer.

Many interferometer applications require an optical arrangement where two interfering beams travel along separate paths before they are recombined. Therefore, a number of interferometers for specific purposes have been developed. Most of them are based on the division of amplitude, with the exception of the Rayleigh interferometer and its variations. Here it will be presented three typical interferometers with separate beam paths and division of amplitude. In order to perform amplitude division one of them uses partially reflecting film made of metal or dielectric commonly known as a beam splitter. Other uses a birefringent element in order to produce two orthogonally polarized beams. In this case both beams must be derived from a single polarized beam and then brought into the same polarization state. The last one uses a scatter plate, which transmittance varies in a random manner, to produce one or more diffracted beams, as well as a reflected or transmitted beam.

There are also two beam interferometers that can have different topologies where the most prominent are the Michelson, Mach–Zehnder, and Sagnac interferometers, which will be analyzed in more details in the following chapters.

The Michelson interferometer

The Michelson interferometer, as shown in figure 4.6., uses a light source S which light is divided by a semireflecting coating deposited on one surface of a plane-parallel glass plate BS (known as a beam splitter) into two beams with approximately equal amplitudes. The beams, which are transmitted and reflected from the beam splitter, are back reflected at two plane mirrors M_1 and M_2 and returned to the beam splitter. After the recombination in the beam splitter they impinge the optical detector D.

The optical path difference between interferometer arms is given by the difference of the optical paths in each arm where the optical path of each arm is given by the sum of the products of the thickness of each medium traversed and its refractive index. So, for the optical path difference we can write:

$$OPD = \sum_i n_i t_i - \sum_j n_j t_j \quad , \qquad (4.29)$$

where n_i is the index of refraction of the i-th component and t_i its thickness in one arm and n_j is the index of refraction of the j-th component and t_j its thickness in the other arm. The interferometer components are made of a material which index of refraction is wavelength dependent. Therefore, the optical path difference will be wavelength dependent. In order to eliminate the influence of the wavelength on the optical path difference it is necessary that both arms contain the same thickness of glass having the same dispersion. The compensating plate C has the role to compensate the influence of the beam splitter BS and has the same thickness and it is made of the same material as the beam splitter. In this way each beam passes three times through the glass material of the beam splitter and the compensating plate.

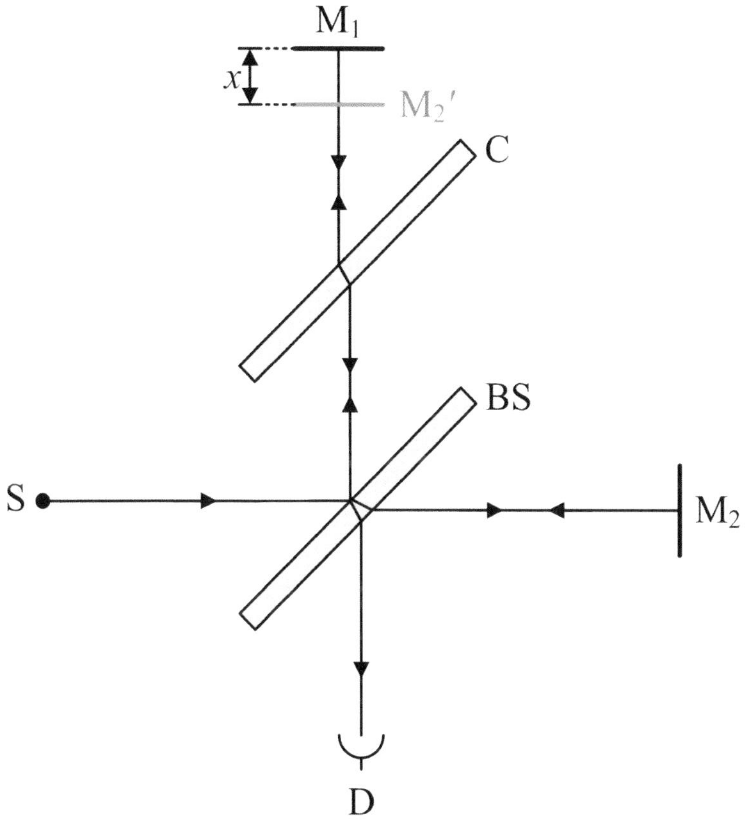

Figure 4.6. The Michelson interferometer.

Reflection at the beam splitter forms an image of mirror M_2 at the position of the virtual mirror M_2'. Therefore, the interference pattern is the same as that formed by a layer of air bounded by the mirror M_1 and M_2', i.e. the optical path difference is equal to twice the distance between them, $OPD = 2x$.

Two virtual point sources (images) S_1 and S_2, which are the images of the monochromatic point source S reflected from mirrors M_1 and M_2, form the nonlocalized fringes on a screen placed at O, as shown in figure 4.7. The case when the beams impinge the mirrors at normal incidence, the line that connects virtual sources S_1 and S_2 is collinear with the beam directed toward the mirror M_2 is shown in figure 4.7.(a). The observed fringes have circular form. If the mirrors are inclined, as presented in figure 4.7.(b), the line that connects virtual sources S_1 and S_2 is normal to the beam directed toward the mirror M_2 and the observed fringes are straight lines.

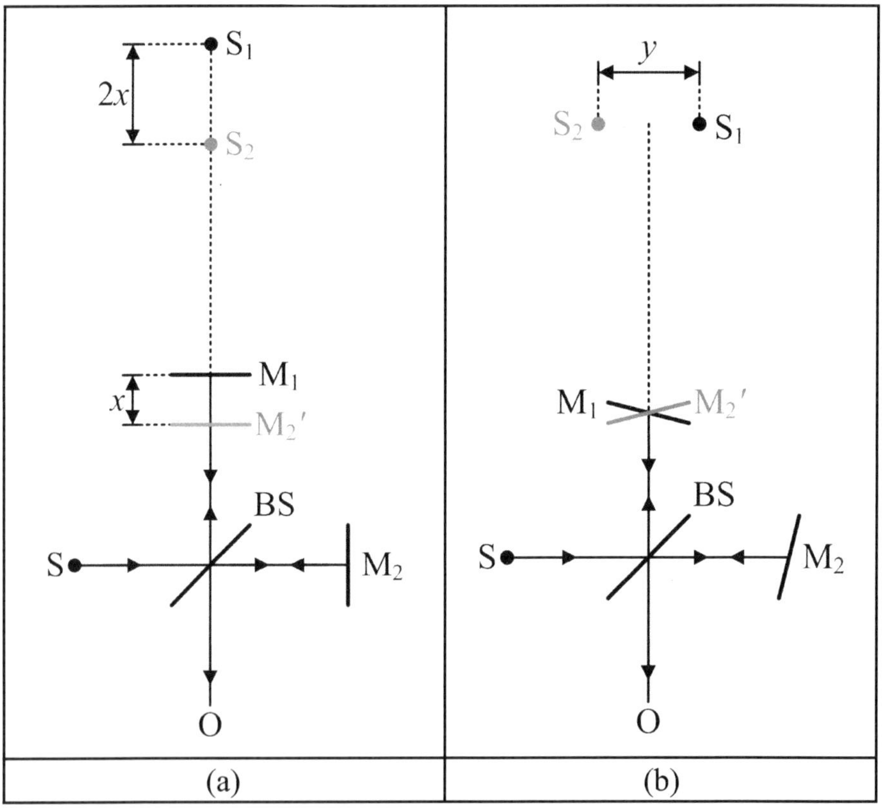

Figure 4.7. Formation of nonlocalized fringes in a Michelson interferometer.

The Mach–Zehnder Interferometer

Figure 4.8. represents the typical setup of the Mach–Zehnder interferometer. Instead of one beam splitter, which have been used in tandem with the compensation plate in the Michelson interferometer, the Mach–Zehnder interferometer uses two beam splitters BS_1 and BS_2 and two mirrors M_1 and M_2. At first place the light from the source S is divided at the semireflecting surface of beam splitter BS_1 into two beams, which, after reflection at the mirrors M_1, M_2, are recombined at the semireflecting surface of the beam splitter BS_2. Typically, beam splitters and mirrors are positioned so that they are approximately parallel, and the paths traversed by the beams form a rectangle or a parallelogram. If it is assumed that the interferometer is illuminated with a collimated beam then we have two plane waves in the interferometer arms. If the wavefront plane P_2' is the image of the wavefront plane P_2 in the beam splitter BS_2 then the phase difference between P_1 and P_2' is then given by:

$$\Delta\varphi = \frac{2\pi}{\lambda} nd \quad , \qquad (4.30)$$

where d is the distance between planes P_1 and P_2' and n is the refractive index of the medium between them.

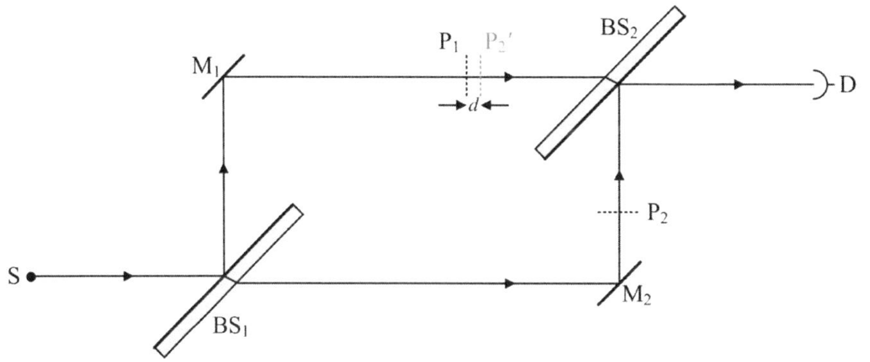

Figure 4.8. The Mach–Zehnder interferometer.

Thanks to its topology, the Mach–Zehnder interferometer is a more versatile optical instrument than the Michelson interferometer because each beam arms are passed over only once, and the fringes can be localized in any plane. Therefore, the Mach–Zehnder interferometer has been widely used in measurement systems where the measurements of the refractive index changes that occur in one arm are related to the changes in pressure, temperature, etc. One of the drawbacks of the Mach–Zehnder interferometer is its adjustment in order to get good fringe visibility with an extended broad-band light source.

The Sagnac Interferometer

The Sagnac interferometer is designed in such a way that the two beams travel around the same closed circuit in opposite directions, as presented in figure 4.9. As the beams travel along the same path the interferometer is extremely stable. Moreover, as the optical paths of the two beams are always very nearly equal it is very easy to align even with a broadband source. Due to the path lengths balance the optical path difference between two beams is equal to zero.

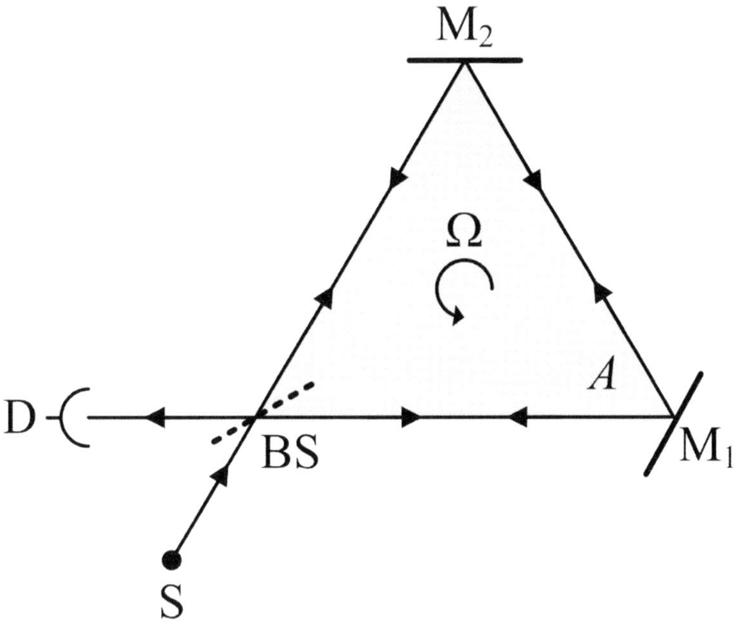

Figure 4.9. The Sagnac interferometer.

In the case when the interferometer setup rotates along the axis perpendicular to the plane of the interferometer a phase shift between the beams occurs, which is given by:

$$\Delta\varphi = \frac{4nA\,\Omega}{\lambda c}, \qquad (4.31)$$

where n is the index of refraction, A is the area of the loop and Ω is the angular rotation frequency.

COHERENCE

Finite size of the spectral bandwidth of a light source requires more powerful theory in order to process the optical phenomena accompanying such sources. The coherence theory, which is basically a statistical theory that describes the properties of the radiation field in terms of the correlation between the vibrations at different points in the field, offers the possibility to describe such phenomena. In the case of a broadband light source the overall electric field can be expressed as:

$$U_r(t) = \int_0^{+\infty} u(\omega)\cos[\omega t + \theta(\omega)]\mathrm{d}\omega \quad , \qquad (4.32)$$

where the electric field $U_r(t)$ has been presented as a sum of the individual monochromatic electric field components, and where $u(\omega)$ and $\theta(\omega)$ are the amplitude and phase, respectively, of a monochromatic component of angular frequency ω. In order to develop a complex representation of the electric field of the broadband light we will introduce the following function:

$$U_i(t) = \int_0^{+\infty} u(\omega)\sin[\omega t + \theta(\omega)]\mathrm{d}\omega \quad . \qquad (4.33)$$

Based on the last two equations the complex electric field can be represented as:

$$\begin{aligned} U(t) &= U_r(t) + jU_i(t) \\ &= \int_0^{+\infty} u(\omega)\exp\{j[\omega t + \theta(\omega)]\}\mathrm{d}\omega \quad , \end{aligned} \qquad (4.34)$$

or, similarly, if $e(\omega)$ is the Fourier transform of the electric field $U_r(t)$ we have:

$$U_r(t) = \frac{1}{2\pi}\int_{-\infty}^{+\infty} e(\omega)\exp(-j\omega t)\mathrm{d}\omega \quad , \qquad (4.35)$$

and consequently:

$$U(t) = \frac{1}{\pi}\int_0^{+\infty} e(\omega)\exp(-j\omega t)\mathrm{d}\omega \quad . \qquad (4.36)$$

It can be easily shown that the following is valid:

$$\langle U(t)U^*(t)\rangle = 2\langle U_r^2(t)\rangle \quad (4.37)$$

and if the factor ½ is ignored the irradiance is given by:

$$I = \langle U(t)U^*(t)\rangle \quad (4.38)$$

Consequently, if the operations on $U_r(t)$ are linear, it is possible to replace it by $U(t)$ and take the real part at the end of the calculation.

The Mutual Coherence Function

The waves originating from different points of a finite size light source or even from the same point of a thermal source exhibits random amplitude and phase fluctuations. If such wave fields illuminate two points in space or the same point at different instants of time exhibit only partial correlation. In order to evaluate such correlation we will consider an optical setup as presented in figure 4.10., which is similar to Young's double slit interference experiment.

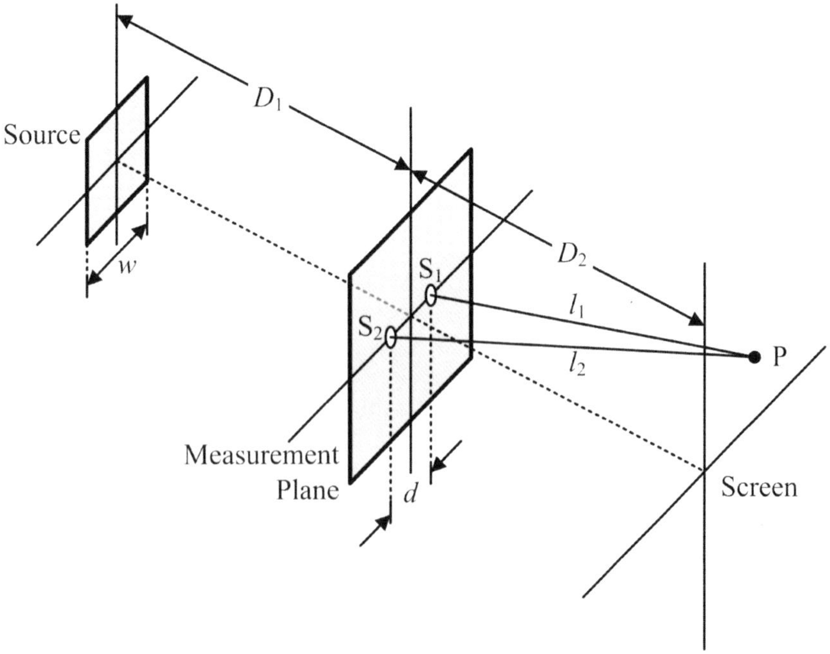

Figure 4.10. Measurement of the coherence of the light produced by a finite size source.

A quasi monochromatic source of finite size illuminates a measurement plane with two pinholes S_1 and S_2, and the light passing through these pinholes forms an interference pattern at a screen. The electric fields caused by the source at the pinholes S_1 and S_2 are $U_1(t)$ and $U_2(t)$, respectively. Therefore, pinholes can been considered as two secondary sources, so that the electric field at a point P on the screen is given by:

$$U_P(t) = \alpha_1 U_1(t - t_1) + \alpha_2 U_2(t - t_2), \tag{4.39}$$

where $t_1 = r_1/c$ and $t_2 = r_2/c$ are the times taken for the waves to travel from the pinholes S_1 and S_2 to travel to the point P on the screen, respectively, and α_1 and α_2 are the scaling factors whose magnitudes are determined by the geometry of the whole setup, such as the size of the pinholes and the distance from the pinholes to the point P. As the interference pattern is stationary, it doesn't depend on the selected time origin, so we can in general case also write:

$$U_P(t) = \alpha_1 U_1(t) + \alpha_2 U_2(t - \tau), \tag{4.40}$$

where τ is the time delay between the beams. For the corresponding irradiance at the point P on the screen we can write:

$$\begin{aligned} I_P &= \langle U_P(t) U_P^*(t) \rangle \\ &= |\alpha_1|^2 \langle U_1(t) U_1^*(t) \rangle + |\alpha_2|^2 \langle U_2(t-\tau) U_2^*(t-\tau) \rangle + \\ &\quad + \alpha_1 \alpha_2^* \langle U_1(t) U_2^*(t-\tau) \rangle + \alpha_1^* \alpha_2 \langle U_1^*(t) U_2(t-\tau) \rangle \\ &= |\alpha_1|^2 I_1 + |\alpha_2|^2 I_2 + 2|\alpha_1 \alpha_2| \mathrm{Re}\{\Gamma_{12}(\tau)\} \end{aligned}, \tag{4.41}$$

where I_1 and I_2 are the irradiances at the pinholes S_1 and S_2, respectively, and

$$\Gamma_{12}(\tau) = \langle U_1(t) U_2^*(t-\tau) \rangle, \tag{4.42}$$

is the mutual coherence function of the wave fields at S_1 and S_2. The mutual coherence function has the same dimensions as intensity. If we normalize it, we obtain the dimensionless quantity

$$\gamma_{12}(\tau) = \frac{\Gamma_{12}(\tau)}{\sqrt{I_1 I_2}}, \tag{4.43}$$

known as the complex degree of coherence of the wave fields at S_1 and S_2. The parameters $|\alpha_1|^2 I_1$ and $|\alpha_2|^2 I_2$ are the irradiances at the point P due to the secondary sources S_1 and S_2 acting separately. Taking the following substitutions $I_{P1} = |\alpha_1|^2 I_1$ and $I_{P2} = |\alpha_2|^2 I_2$ in the equations (4.41) and (4.43) we have:

$$I_P = I_{P1} + I_{P2} + 2\sqrt{I_{P1} I_{P2}}\, \mathrm{Re}\{\gamma_{12}(\tau)\}, \qquad (4.44)$$

which represents the general law of interference for partially coherent light. The complex degree of coherence can be represented as the product of its modulus and its phase factor, so we have:

$$\gamma_{12}(\tau) = |\gamma_{12}(\tau)| \exp\{j[\varphi_{12}(\tau) - \bar{\omega}\tau]\}, \qquad (4.45)$$

where $\varphi_{12}(\tau)$ is the phase difference between the waves incident at S_1 and S_2, and $\bar{\omega}$ is the mean angular frequency. The parameter $|\gamma_{12}(\tau)|$ is known as the degree of coherence. According to the equation (4.41) and the Cauchy-Schwarz-Bunyakovsky inequality we have the following condition fulfilled:

$$0 \le |\gamma_{12}(\tau)| \le 1. \qquad (4.46)$$

If it is satisfied $|\gamma_{12}(\tau)| = 1$ the wave fields at S_1 and S_2 are coherent, and if it is satisfied $|\gamma_{12}(\tau)| = 0$ the wave fields at S_1 and S_2 are incoherent. In all other cases they are partially coherent. The combination of equations (4.44) and (4.45) gives:

$$I_P = I_{P1} + I_{P2} + 2\sqrt{I_{P1} I_{P2}}\, |\gamma_{12}(\tau)| \cos[\varphi_{12}(\tau) - \bar{\omega}\tau], \qquad (4.47)$$

Typically, in the case of the quasi-monochromatic light, the parameters $|\gamma_{12}(\tau)|$ and $\varphi_{12}(\tau)$ slowly varying with respect to the time delay (τ) variation when compared with $\bar{\omega}\tau$, i.e. $\bar{\omega} > \mathrm{d}|\gamma_{12}(\tau)|/\mathrm{d}\tau, \mathrm{d}\varphi_{12}/\mathrm{d}\tau$ is fulfilled. According to equation (4.47) the fringe visibility is given by:

$$V = \frac{2\sqrt{I_{P1} I_{P2}}}{I_{P1} + I_{P2}} |\gamma_{12}(\tau)|, \qquad (4.48)$$

which in the case of two equally intense beams reduces to:

$$V = |\gamma_{12}(\tau)|. \tag{4.49}$$

Spatial Coherence

In order to evaluate the spatial coherence of the wave field, we will consider two points P_1 and P_2 located at the screen, as shown in figure 4.11., illuminated by a small source *S*, whose dimensions are negligible if compared to its distance from the points P_1 and P_2. Further, we will assume that $\Delta U_1(t)$ and $\Delta U_2(t)$ are the elementary electric fields at the points P_1 and P_2 caused by the small radiating element ΔS positioned at the surface of small source S at the position (x,y), the overall electric fields originating from the whole source *S* at the points P_1 and P_2 are given by:

$$\begin{aligned} U_1(t) &= \sum_S \Delta U_1(t) \\ U_2(t) &= \sum_S \Delta U_2(t) \end{aligned} \tag{4.50}$$

The mutual irradiance is given by:

$$\begin{aligned} I_{12} &= \langle U_1(t) U_2^*(t) \rangle \\ &= \sum_S \langle \Delta U_1(t) \Delta U_2^*(t) \rangle \end{aligned}, \tag{4.51}$$

where due to the independent fluctuations of the wave fields, originating from different elementary areas of the source, the sum elements in equation (4.52) that represent the time average of such fields are omitted. The elementary electric fields at the points P_1 and P_2 can be represented as:

$$\begin{aligned} \Delta U_1(t) &= \frac{\Delta u\left(t - \frac{l_1}{c}\right)}{l_1} \exp\left[-j\overline{\omega}\left(t - \frac{l_1}{c}\right)\right] \\ \Delta U_2(t) &= \frac{\Delta u\left(t - \frac{l_2}{c}\right)}{l_2} \exp\left[-j\overline{\omega}\left(t - \frac{l_2}{c}\right)\right], \end{aligned} \tag{4.52}$$

where $\Delta u(t)$ is the complex amplitude of the wave field at the element ΔS, and l_1 and l_2 are the distances from ΔS to the points to P_1 and P_2, respectively. Consequently we have:

$$\Delta U_1(t)\Delta U_2^*(t)=\frac{\Delta u\left(t-\frac{l_1}{c}\right)\Delta u^*\left(t-\frac{l_2}{c}\right)}{l_1 l_2}\exp\left[-j\overline{\omega}\left(\frac{l_2-l_1}{c}\right)\right]. \quad (4.53)$$

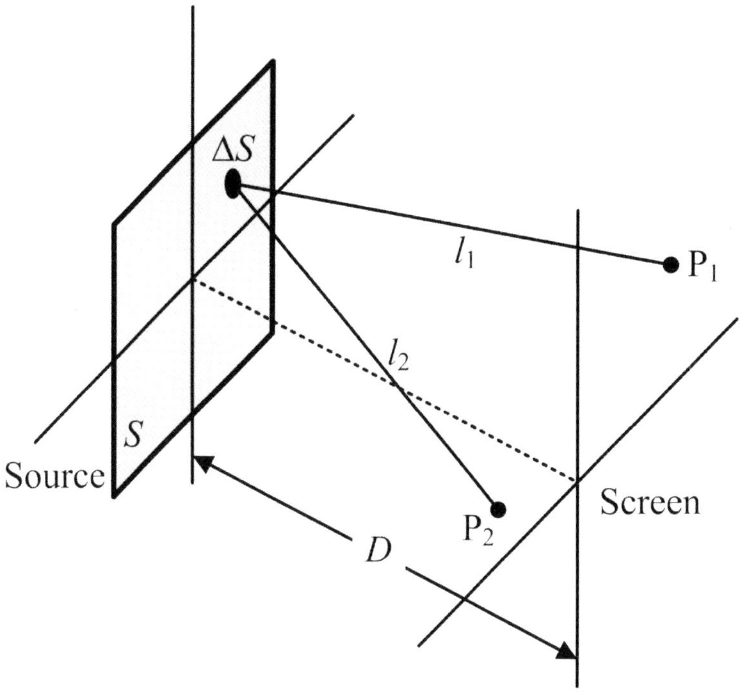

Figure 4.11. Spatial coherence of the fields at two points P_1 and P_2 illuminated by an extended source.

With a quasi-monochromatic source, which emits only in a narrow range of frequencies of $\overline{\nu}\pm\Delta\nu/2$, the complex amplitude $\Delta u(t)$ varies slowly enough, thus, in the case where time interval $(l_2-l_1)/c$ is much smaller if compared with the time interval $1/\Delta\nu$ we have $\Delta u(t-l_1/c)\approx\Delta u(t-l_2/c)\approx\Delta u(t)$ and consequently:

$$\Delta U_1(t)\Delta U_2^*(t)=\frac{|\Delta u(t)|^2}{l_1 l_2}\exp\left[-j\overline{\omega}\left(\frac{l_2-l_1}{c}\right)\right]. \quad (4.54)$$

In order to obtain the overall mutual irradiance emerging from the whole source the last equation must be integrated over the whole area of the source, so we have:

$$I_{12} = \int_S \frac{I(x,y)}{l_1 l_2} \exp\left[-j\bar{k}(l_2 - l_1)\right] dS \quad , \qquad (4.55)$$

where it was taken into account the following relation:

$$I(x,y)dS = |\Delta u(t)|^2 \quad , \qquad (4.56)$$

and $\bar{k} = \bar{\omega}/c = 2\pi/\bar{\lambda}$. If I_1 and I_2 are the irradiances at the points P_1 and P_2 due to the whole source S, respectively we similarly have:

$$I_1 = \int_S \frac{I(x,y)}{l_1^2} dS$$

$$I_2 = \int_S \frac{I(x,y)}{l_2^2} dS \quad . \qquad (4.57)$$

Finally, for the complex degree of coherence of the fields at the points P_1 and P_2 we obtain:

$$\mu_{12} = \frac{1}{\sqrt{I_1 I_2}} \int_S \frac{I(x,y)}{l_1 l_2} \exp\left[-j\bar{k}(l_2 - l_1)\right] dS \quad . \qquad (4.58)$$

If the dimensions of the source, and the separation of P_1 and P_2, are much smaller than the distances of P_1 and P_2 from the source, then the last equation can be simplified.

Coherence Time and Coherence Length

In the case of a source having very small dimensions, i.e. in the case of a point like source, which radiates over a range of wavelengths, we have to deal with the temporal coherence of the field. Therefore, the complex degree of coherence depends only on the difference in the transit times from the source to P_1 and P_2, and the mutual coherence function, as a matter of fact, represents the autocorrelation function:

$$\Gamma_1(\tau) = \langle U(t)U^*(t-\tau)\rangle \quad , \qquad (4.59)$$

and the degree of temporal coherence of the field is then given by:

$$\gamma_1(\tau) = \frac{\langle U(t)U^*(t-\tau)\rangle}{\langle U(t)U^*(t)\rangle} \qquad (4.60)$$

The degree of temporal coherence can be obtained from the visibility of the interference fringes as the optical path difference varies.

According to the Wiener-Khinchin theorem the radiation power spectrum is given by the Fourier transform of the mutual coherence function (autocorrelation function), so that the mutual coherence function and the radiation spectrum represent the Fourier transform pair:

$$S(\nu) = F\{\Gamma_1(\tau)\}$$
$$\Gamma_1(\tau) = F^{-1}\{S(\nu)\} \qquad (4.61)$$

where $S(\nu)$ is the radiation power spectrum. Consequently, the complex degree of coherence is given by:

$$\gamma_1(\tau) = \frac{F^{-1}\{S(\nu)\}}{\int_{-\infty}^{+\infty} S(\nu)\,\mathrm{d}\nu} \qquad (4.62)$$

If we now consider the radiation source that emits the radiation evenly distributed over the frequency range of $\bar{\nu} - \Delta\nu/2 \leq \nu \leq \bar{\nu} + \Delta\nu/2$, i.e. centered at the mean frequency of $\bar{\nu}$ and with the spectrum width equal to $\Delta\nu$, then according to the equation (4.62) the complex degree of coherence of the radiation is given by:

$$\gamma_1(\tau) = \frac{\sin(\Delta\nu\tau)}{\Delta\nu\tau} \qquad (4.63)$$

The sinc function is a damped oscillating function whose first zero occurs at a time difference τ_C given by:

$$\Delta\nu\tau_C = 1 \qquad (4.64)$$

The time interval τ_C is also known as the coherence time of the radiation. Consequently, the coherence length is defined as:

$$l_C = c\tau_C = \frac{c}{\Delta\nu} \approx \frac{\bar{\lambda}^2}{\Delta\lambda} \qquad (4.65)$$

On a clear, dark night, our eyes can see about 6,000 or so stars in the sky. One of the most interesting phenomena of the clear night sky is why some stars twinkling and some not. In order to determine which kind of celestial bodies twinkle and which not we will observe a star with the radius R located at a very large distance D from the observer. Two rays, which emerge from the point P located at the star surface, travel along two paths with the distance l_1 and l_2 until reach the observer eye. After the refraction at the eye lens the rays impinge the retina at the point R. Due to the different ray paths the turbulence of the atmosphere can cause random phase modulation of the wave fields which further lead to the random fluctuations at the observer retina known as the star twinkling. According to the geometry from figure 4.12., we have:

$$l_1 = \sqrt{D^2 + \left(R - \frac{d}{2}\right)^2} \approx D\left(1 - \frac{1}{2}\frac{Rd}{D^2}\right)$$

$$l_2 = \sqrt{D^2 + \left(R + \frac{d}{2}\right)^2} \approx D\left(1 + \frac{1}{2}\frac{Rd}{D^2}\right), \quad (4.66)$$

where d is the pupil diameter and where the following condition $d < R < D$ were taken into consideration.

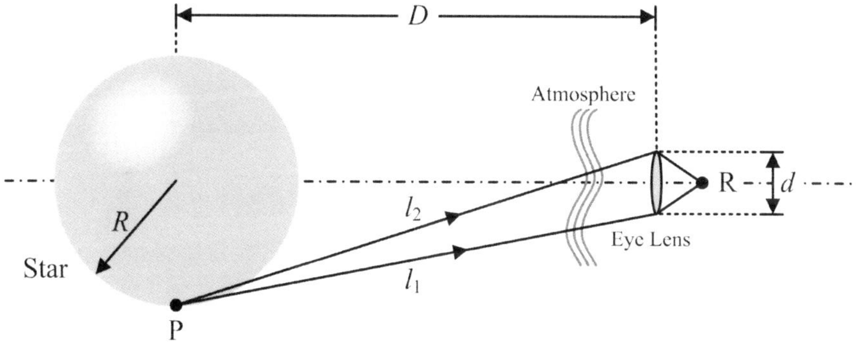

Figure 4.12. Star twinkle.

The optical path difference of these two rays is given by:

$$\Delta l = l_2 - l_1 \approx \frac{Rd}{D}. \quad (4.67)$$

Roughly speaking the star will twinkle if the optical path difference is smaller than the coherence length ($\Delta l < l_C$) and will not if the optical path difference is larger than the coherence length ($\Delta l > l_C$). Therefore, if the star apparent size, defined as:

$$\varphi \approx \frac{2R}{D}, \qquad (4.68)$$

is smaller than:

$$\varphi < \frac{2l_C}{d}, \qquad (4.69)$$

the star will twinkle and in opposite will not. The coherence length of the light emitted by the star depends on the star temperature (black body radiation) and the atmosphere absorption. Therefore, the coherence length has not the fixed value but depends on the particular star characteristics. However, regardless the star light spectrum, our eyes are sensitive only in a certain wavelength range. According to the spectral luminous efficiency function for photopic vision an average human eye has a spectral bandwidth ranging from $\lambda_L \approx 380$ nm to $\lambda_H \approx 770$ nm, thus having a spectral width of $\Delta\lambda = \lambda_H - \lambda_L \approx 390$ nm, and the peak sensitivity at $\bar{\lambda} = 555$ nm. Therefore, the effective coherence length, given by equation (4.65), is equal to $l_C \approx 0.8$ μm. As the observer looks at the sky during the night, the pupil has the maximal openings of approximately $d \approx 8$ mm. So, according to equation (4.69) if the apparent size is smaller than 200 μrad or 40″ the star will twinkle and in opposite will not or it twinkling will be weaker. If we compare this approximate limit value for the star apparent size and the apparent sizes of the particular celestial bodies, shown in table 4.1., one can see why stars twinkle distinctively from the planets which in some case may also twinkle but not so noticeable.

Table 4.1. Apparent sizes of some celestial bodies

	Planets			Stars		
Celestial body	Venus	Mars	Jupiter	Sirius	Alpha Centauri	Betelgeuse
Apparent size	9.6″ - 66″	3.5″ - 25″	30″ - 50″	0.006″	0.007″	0.05″ - 0.06″

WHITE-LIGHT INTERFEROMETRY

If the interferometer has been illuminated with a monochromatic light one can notice a distinctive interferometric pattern on the observation screen or if an optical detector has been used in order to capture the fringes when the optical path difference in the interferometer changes one can notice a very large number of fringes, which in the case of a linear optical path difference sweep produces a periodic fringe pattern. In theory we will have an unlimited number of fringes. However, typical light source has a finite spectrum width thus limiting the number of observed/detected fringes. According to the Wiener-Khinchin theorem the broader the spectrum the shorter the interferometric pattern. One of the most common light sources used in the white-light interferometer or also called low-coherence interferometer is the superlumiescent diode. Typical superluminescent diode emits infrared light, which has Gaussian spectrum shape given as:

$$S(\nu)=\frac{2}{\Delta\nu}\sqrt{\frac{\mathrm{h}\ 2}{\pi}}\exp\left[-4\mathrm{h}\ 2\left(\frac{\nu-\bar{\nu}}{\Delta\nu}\right)^2\right], \qquad (4.70)$$

where $\bar{\nu}$ is the central frequency and $\Delta\nu$ is the full width at half maximum (FWHM) spectrum width. The complex degree of coherence of the radiation is he given by:

$$|\gamma_1(\tau)|=\exp\left[-\left(\frac{\pi\Delta\nu}{2\sqrt{\mathrm{h}\ 2}}\tau\right)^2\right]. \qquad (4.71)$$

Consequently, the FWHM coherence time and the coherence length are:

$$\tau_C=\frac{\sqrt{2\mathrm{h}\ 2}}{\pi}\frac{1}{\Delta\nu}$$

$$l_C=\frac{\sqrt{2\mathrm{h}\ 2}}{\pi}\frac{\bar{\lambda}^2}{\Delta\lambda}. \qquad (4.72)$$

Due to the very short wavelength of the light, interferometry offers the possibility of measuring very small distances. However, the periodicity of the intrefreometric pattern, when the monochromatic source has been used, limits the dynamic range of the measurement. The

2π ambiguity in interferometer phase unwrapping can limit the maximal distance measurement to the fraction of the wavelength. Fringe counting is one of the possible solutions in measurement larger distances with high-coherence interferometry, which in general case requires moving optical elements that can influence measurement precision and accuracy. Where the classical high-coherence interferomerty fails, the white-light interferometry manages to measure larger distances even without moving parts. As the interferometer optical path difference changes with the changes of a distance of some interferometer components the center of the complex degree of coherence of the radiation also proportionally changes. Therefore, by extracting the position of the maximum value of the complex degree of coherence one can absolutely measure the interferometer optical path difference and thus the measuring object position, as presented in figure 4.12. where the white-light Twyman-Green interferometer is presented.

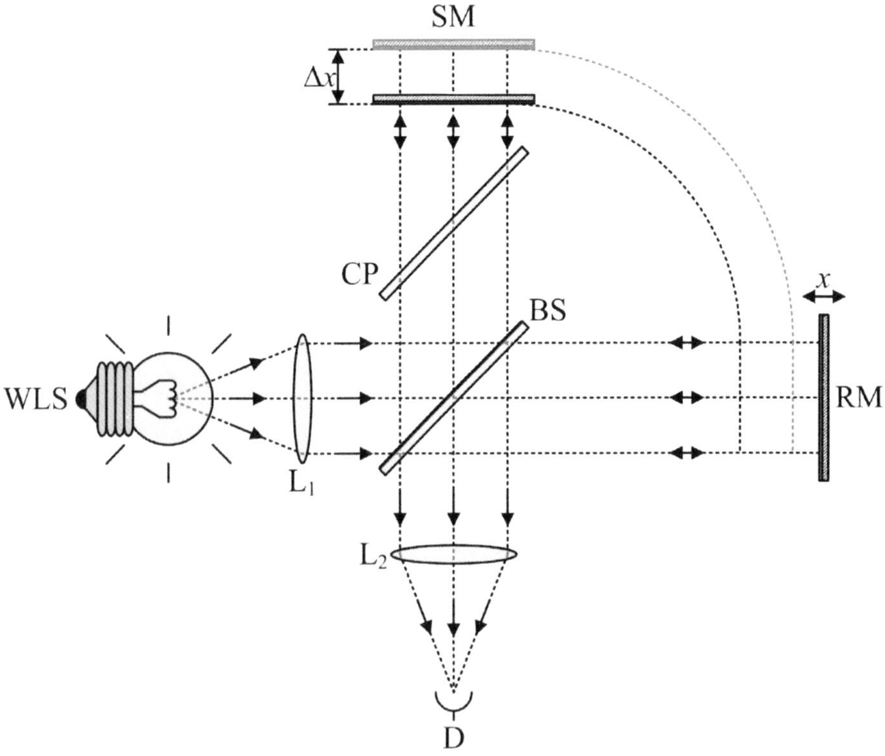

Figure 4.13. White-light Twyman-Green interferometer.

When the sensor mirror SM, which is fixed to an object, which position is to be measured, moves along one of the interferometer arm in order to measure its position the reference mirror RM must perform scanning along the other interferometer arm. To be able to measure the absolute position of SM the scanning mechanism of the reference mirror must be equipped with an internal RM position measurement during the scanning process. The scanning of RM will give the interferometric patterns as presented in figure 4.14.

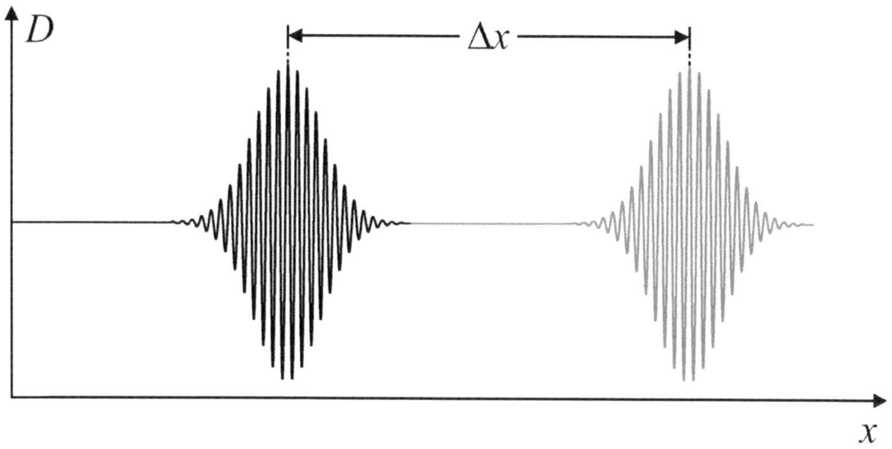

Figure 4.14. White-light interferometric patterns for two different optical path differences.

According to figure 4.14. if SM is displaced for the distance Δx the maximum of the of the complex degree of coherence will be also shifted along the x axis for the same distance. The drawback of such a setup is the movable parts, which can influence the measurement. The movable parts can be omitted if the electronic scanning of the white-light interferometeric pattern is employed as presented in figure 4.15. Instead of moving the RM an optical wedge in tandem with CCD linear sensor can be used. The optical wedge with a small apex angle θ is made of a glass with the refractive index n, which planes are partially reflective, as presented in figure 4.16. Due to the linear increase of the optical path difference in the wedge along the CCD linear array a low-coherence interferometric pattern emerges. The corresponding optical path difference at the distance x from the wedge apex is given by:

$$OPD = 2n\theta x \,. \tag{4.73}$$

The maximum or the center of gravity of the complex degree of coherence will be positioned at the CCD linear array pixel which position x_C is determined according to the optical paths difference balance in the sensing interferometer and the receiving interferometer made of the wedge, so we have:

$$OPD = 2n\theta x_C = 2\Delta x$$

$$x_C = \frac{\Delta x}{n\theta} \quad , \qquad (4.74)$$

thus giving the linear relation between the interferogram center position and the absolute position of the sensing mirror.

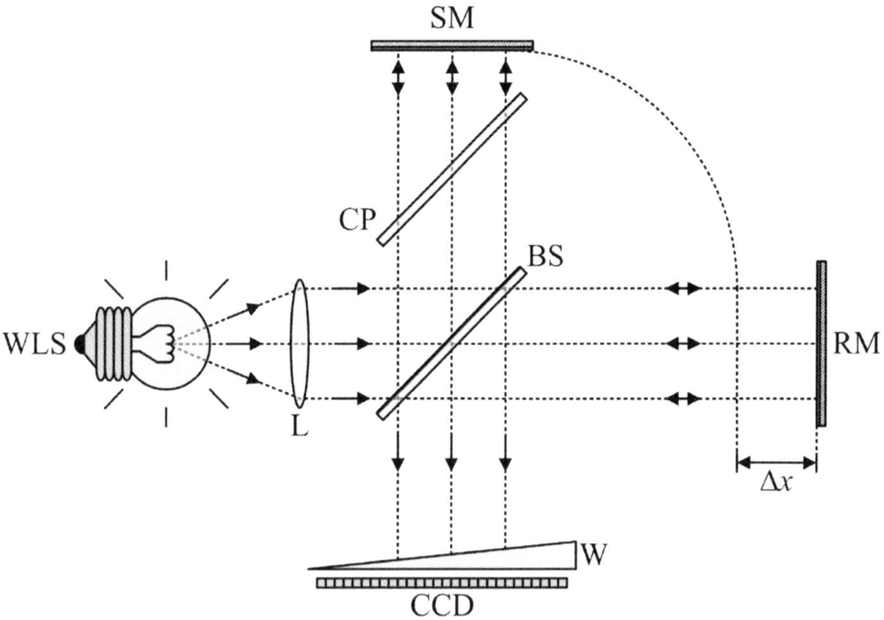

Figure 4.15. White-light Twyman-Green interferometer with electronic scanning.

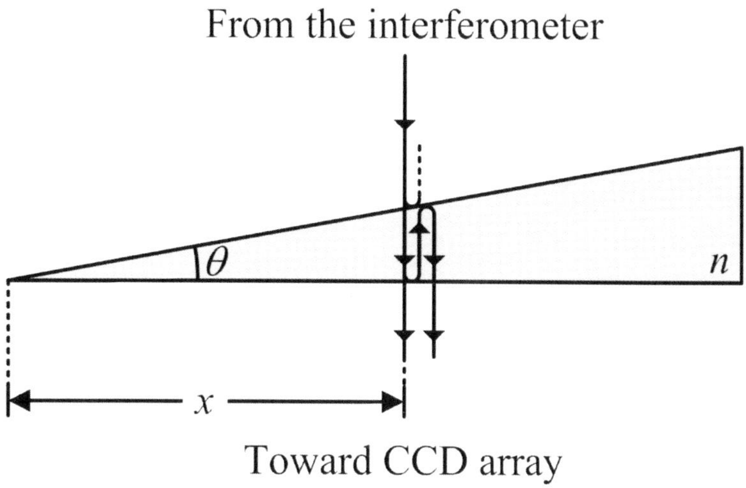

Figure 4.16. Optical wedge for electronic scanning.

If instead of an optical detector such as photodiode or CCD linear array, a dispersive element has been employed one can obtain the spectral fringes or the so called channeled spectrum. The optical setup for obtaining the channeled spectrum is presented in figure 4.17. The light, obtained from a low-coherence interferometer, illuminates the slit S where after passing through the slit the broadband light illuminates an optical spectrometer OS. The radiation that illuminates the spectrometer is consisted of two wave fields:

$$U_{\otimes}(t) = aU(t) + bU(t - \Delta t), \qquad (4.75)$$

i.e. the superposition of the reference beam field $U(t)$ and the same field delayed by propagating the path length difference $2\Delta x$ $U(t\text{-}\Delta t)$ where $\Delta t = 2\Delta x/c$ is fulfilled and where the parameters a and b are inserted in the equation in order to account the possible differences in relative amplitudes between the reference and delayed fields. The spectrum of the wave field captured by the optical spectrometer is found from the Fourier transform of equation (4.75) as:

$$S(\nu) = \left(a^2 + b^2\right) WLS(\nu)\left[1 + V\cos\left(\frac{4\pi\Delta x}{c}\nu\right)\right], \qquad (4.76)$$

where $WLS(\nu)$ is the spectrum of the white-light source WLS and V is the spectral fringe visibility given by:

$$V = \frac{2b}{a^2 + b^2}. \quad (4.77)$$

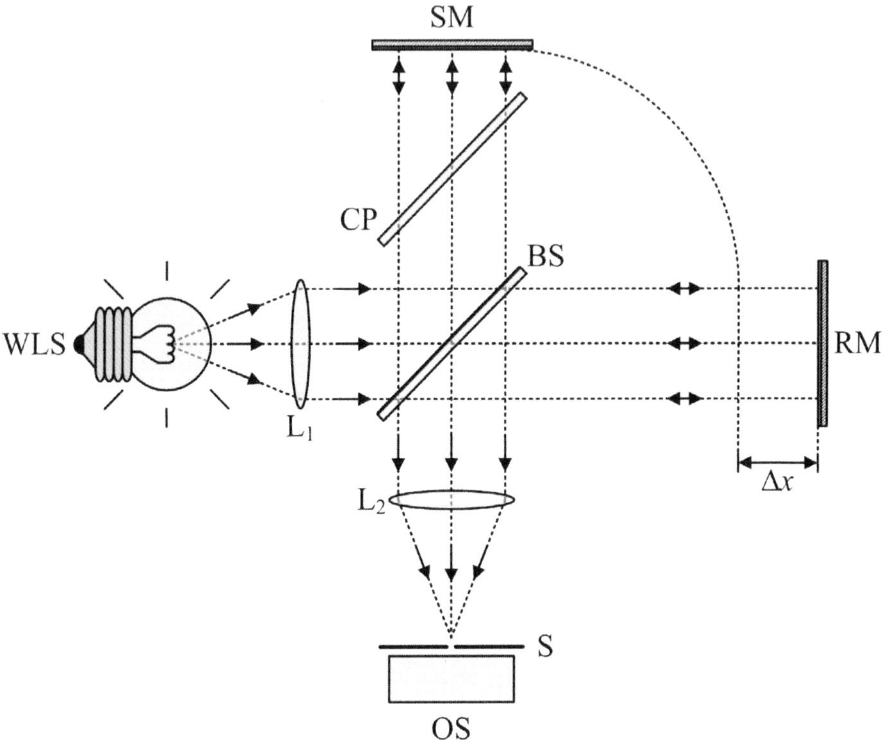

Figure 4.17. Channeled spectrum.

The typical shape of the channeled spectrum is presented in figure 4.18. where are shown two spectral fringes for $\Delta x = 90$ μm and $\Delta x = 110$ μm where as a broadband light source a superluminescent diode is used. The diode has a central wavelength of 850 nm and spectrum width of approximately 20 nm, which further results in the coherence length of the diode of about 36 μm.

Spectral interferometry is a very powerful interrogation tool especially in the case of a large number of quasi-distributed extrinsic low-finesse Fabry-Pérot interferometric sensors aimed, for example, for distributed measurements of temperature, pressure, force, strain, etc. This technique allows us multiplexing a large number of sensors placed along a single optical fiber. As shown in figure 4.18., a white-light source spectrum is modulated, where the modulation "frequency" depends on the optical path difference. So, in the case of several sensing interferometers we will have more complex modulation scheme with the same number as the sensor number of modulation "frequencies". Therefore, by locking on a single "frequency" of the captured channeled spectrum one can track the changes in "frequency" of the particular sensor and perform the necessary measurements on this particular sensor.

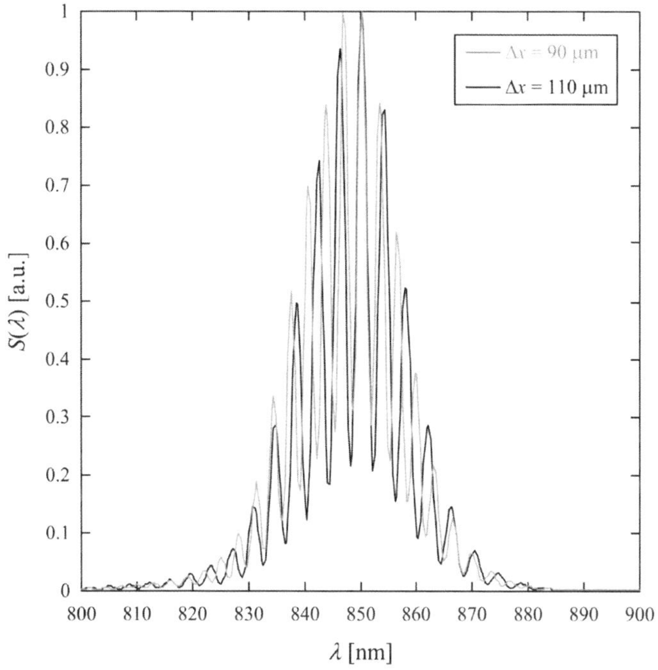

Figure 4.18. Measured channeled spectrum.

Multiplexing capabilities of the white-light interferometry, i.e. the simultaneous measurement of several optical path differences, are used in the optical coherence-domain reflectometry that is suitable for determining the positions and magnitudes of reflection sites within a miniature multilayered optical assemblies, as presented in figure 4.19. Typical optical coherence reflectometer / tomography measurement system captures the internal reflections from the structured miniature object. The internal reflections are caused by the scattering of the incident optical radiation on a nonhomogeneous internal structure of the object. In order to investigate the positions, sizes and partial characteristics of the layers of such internal structured object, the reference mirror performs scanning where at the detector D are captured the corresponding low-coherence interferometric patterns. These patterns are presented in figure 4.20. where each pattern corresponds to a single reflection. Position of each layer corresponds to the interferogram center position and where the reflection coefficient can be determined from the interferogram amplitude. One can notice that the resolution of the optical coherence tomography is approximately equal to the coherence length of the WLS. Typical optical coherence tomography systems have the resolution of about 10 μm.

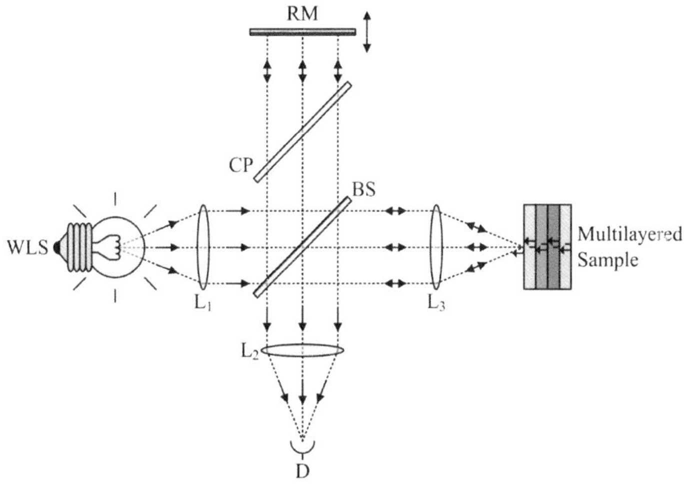

Figure 4.19. Optical coherence-domain reflectometry.

Instead of mechanical scanning and a simple optical detector one can use the fixed reference mirror in tandem with the optical spectrometer. However, such a system suffers from lower signal-to-noise ratio then one with the movable reference mirror and single optical detector, which can be the limiting factor in the case of weak scattering.

One of the most prominent applications of the optical coherence tomography is in medical imaging in order to obtain micrometer-resolution, three-dimensional images of biological tissues. Depending on the properties of the used white-light source such as superluminescent diodes, ultrashort pulsed lasers, and supercontinuum lasers, optical coherence tomography has achieved sub-micrometer resolution typically with a very wide-spectrum sources.

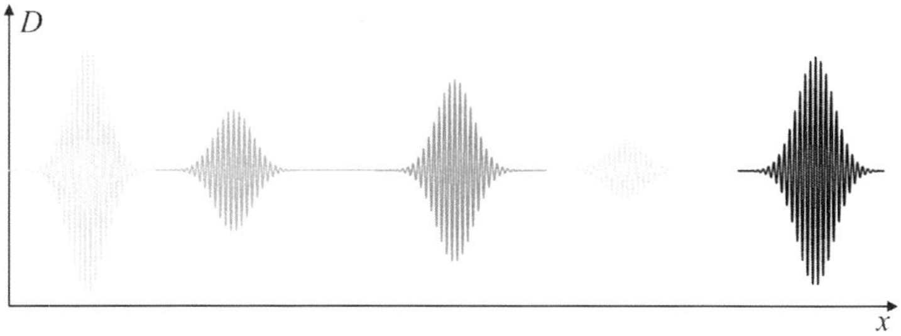

Figure 4.20. Captured optical coherence tomography interferometric patterns where each pattern corresponds to a single reflection.

White-light interferometry represents also a powerful tool for measuring a surface's profile, in order to quantify its roughness as presented in figure 4.21. A light that is back reflected from the rough surface is captured by a fast CCD two dimensional array (camera). In order to perform surface profiling the reference mirror is moved typically with the help of a piezo actuator. The camera captures images where the signal at each camera pixel represents the low-coherence interferometric pattern obtained from a particular point at the rough surface. By finding the maximum of the interferometric pattern one can obtain the position of this particular point with respect to some reference plane. By analyzing in such way each interferogram obtained from all the camera pixels one can obtain the three dimensional image of the surface.

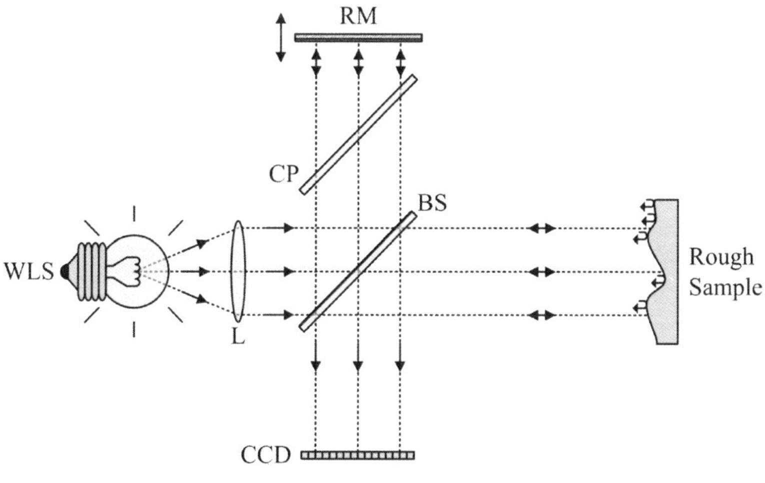

Figure 4.21. White-light profilometry.

MULTIPLE BEAM INTERFERENCE

In the previous discussion in this chapter regarding the interferometric phenomena occurring in the plane-parallel plate, we have taken into account only two reflected beams and ignored any multiple reflections. If the slab has low refection coefficients at its planes we can consider this interferometer to be low-finesse, i.e. we can consider this interferometer as a two beam interferometer. However, if special measures are taken in order to increase the reflection coefficients so that they are enhanced, allowing more beams to be combined, the fringes change their character and become very much sharper. If a thin metallic film is deposited onto the slab surfaces it is possible to increase the reflection coefficient. In this way the slab front face reflects a large fraction of the light incident upon it. The small amount of light that passes through to the inside of the slab is reflected back and forth many times. Therefore, we have the interference of these many emerging rays, which gives multiple interference its special character.

The case of simple multiple interference is presented in figure 4.22. Light source S emits the light toward the plane-parallel plate of thickness d and index of refraction n. Due to the multiple reflections we have reflected rays R_1, R_2, R_3, ... as well as transmitted rays T_1, T_2, T_3, ..., where each ray with the larger index is weaker than one with the smaller index due to the larger number of internal reflections and thus larger losses. If the amplitude of the incident ray is U and r and t the reflection and transmission coefficients from the surrounding medium to the slab, respectively and r' and t' the corresponding quantities from slab to the surrounding medium, we have the following for the amplitude of the reflected rays:

$$U_{R_1} = rU$$
$$U_{R_i} = tU\,t'r'^{2i-1}, \quad i > 1 \quad , \qquad (4.78)$$

and of the transmitted rays:

$$U_{T_i} = tU\,t'r'^{2(i-1)} \quad . \qquad (4.79)$$

With the exception of the first reflected ray R_1 all other reflected ray amplitude follows the geometric progression. Each ray phase depends on the angle of refraction inside the plate, its thickness and its refraction index. The relative phase of the rays on a plane outside the slab perpendicular to the ray depends as:

$$\varphi = \frac{4\pi nd \cos\theta_2}{\lambda} \qquad (4.80)$$

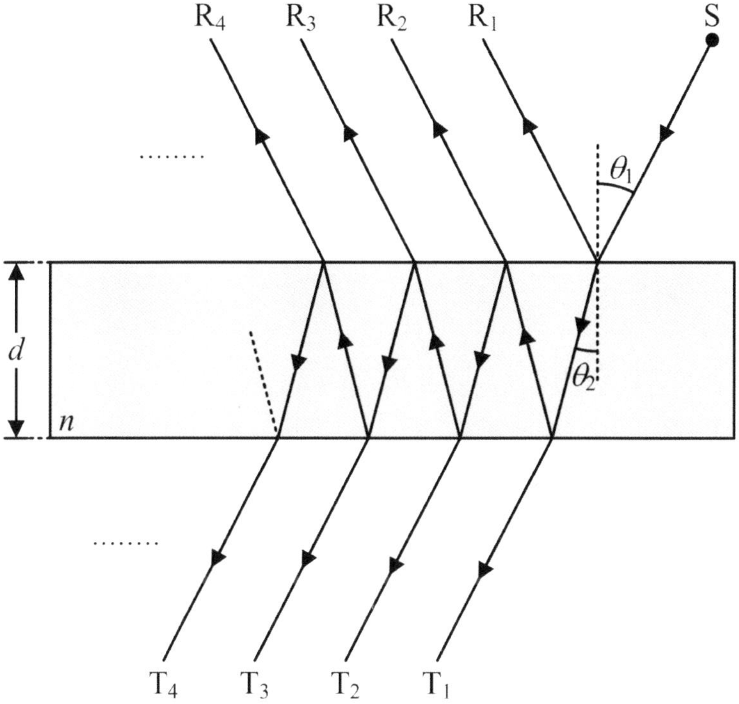

Figure 4.22. Multiple-beam interference in a plane-parallel plate.

In the case of the overall transmitted beam amplitude we have the following:

$$U_{\mathrm{T}} = \sum_{i=1}^{+\infty} U_{\mathrm{T}_i} = \sum_{i=1}^{+\infty} U\, t' r'^{2(i-1)} \exp[-j(i-1)\varphi] , \qquad (4.81)$$

where after infinite summation the following is obtained:

$$U_{\mathrm{T}} = U\, t' \frac{1}{1 - r'^2 \exp(-j\varphi)} . \qquad (4.82)$$

Bearing in mind that the transmitted irradiance is given by $I_{\mathrm{T}} = U_{\mathrm{T}} U_{\mathrm{T}}^*$, we have:

$$I_T = \frac{|U\,t'|^2}{\left[1 - r'^2 \exp(-j\varphi)\right]\left[1 - \left(r'^2\right)^* \exp(j\varphi)\right]}$$

$$I_T = \frac{|U\,t'|^2}{1 - 2\mathrm{R}\left\{r'^2 \exp(-j\varphi)\right\} + \left|r'^2\right|^2}\ . \qquad (4.83)$$

As for the incident wave irradiance is valid $I_I = |U|^2$, equation (4.83) becomes:

$$\frac{I_T}{I_I} = \frac{|tt'|^2}{1 - 2\mathrm{R}\left\{r'^2 \exp(-j\varphi)\right\} + \left|r'^2\right|^2}\ . \qquad (4.84)$$

In general case, the reflection coefficients from to the slab to the surrounding medium is a complex quantity where is valid $r' = |r'|\exp(j\phi)$, where ϕ is the phase change due to the reflection so, equation (4.84) becomes:

$$\frac{I_T}{I_I} = \frac{|tt'|^2}{1 - 2|r'|^2 \cos\psi + |r'|^4}, \qquad (4.85)$$

where $\psi = 2\phi - \varphi$ is valid. Based on the energy conservation low we have fulfilled $|tt'|^2 = 1 - |r'|^2$, so equation (4.85) becomes:

$$\frac{I_T}{I_I} = \frac{1 - |r'|^2}{1 - 2|r'|^2 \cos\psi + |r'|^4}, \qquad (4.86)$$

where after the rearrangement of equation (4.86) we have:

$$\frac{I_T}{I_I} = \frac{1}{1 + F\sin^2\left(\frac{\psi}{2}\right)}, \qquad (4.87)$$

and:

$$F = \frac{2|r'|^2}{\left(1 - |r'|^2\right)^2}\ . \qquad (4.88)$$

One can notice from the last equation that if parameter $|r'|^2$ approaches unity the parameter F becomes very large. For example, if $|r'|^2 = 0.9$, $F = 180$. The consequence of this is to keep the transmitted irradiance very small for a broad ranges of wavelengths, except when ψ is close enough to a $2k\pi$, where k is integer, where $\sin^2(\psi/2)$ becomes less than $1/F$. The transmission coefficient I_T/I_I plots for five different values of the parameter F are given in figure 4.23. The sharp fringes of multiple beam interference are known as Fabry-Pérot fringes.

The Fabry-Pérot fringes sharpness is often itemized as the finesse Φ, which is determined as the ratio of the separation of adjacent maximums to the fringe half-width. Equation (4.87) gives the phase shift $\psi_{1/2}$ for the fringe irradiance to be halved as:

$$1 + F\sin^2\left(\frac{\psi_{1/2}}{2}\right) = 2, \tag{4.89}$$

or:

$$\psi_{1/2} = 2\arcsin\left(\frac{1}{\sqrt{F}}\right). \tag{4.90}$$

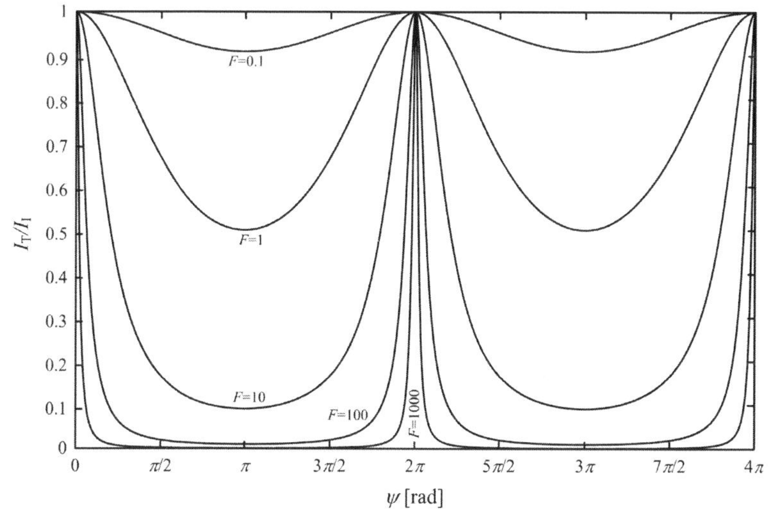

Figure 4.23. Fabry-Pérot fringes for five values of the parameter F.

Typically parameter F is large, therefore we have $\arcsin\left(1/\sqrt{F}\right) \approx 1/\sqrt{F}$. Since the phase difference between two adjacent fringes is 2π, the finesse Φ is the ratio $2\pi/2\psi_{1/2}$:

$$\Phi = \frac{2\pi}{2\psi_{1/2}} \approx \frac{\pi\sqrt{F}}{2} \quad (4.91)$$

The Fabry-Pérot interferometer, presented in figure 4.24., uses the previously discussed characteristic of such a multiple beams interferometer in order to obtain sharp interference fringes from the source S. The interference pattern, observed on the plate P, is in the shape of concentric rings, which are the images of those points on the source produced with the help of the first lens. Usually, the cavity is made of two glass plates. These plates have their inner surfaces coated with the films of high reflectivity. These parallel plates are held at a certain distance by an optical spacer made of invar or silica. This device is called a Fabry-Pérot etalon.

When the Fabry-Pérot etalon is illuminated with the help of a light containing two spectral components one can observe double ring system. With the etalon it is possible to optically distinguish hyperfine structure of spectral lines directly. Therefore, the Fabry-Pérot etalon since its introduction dominate the field of high-resolution spectrometry.

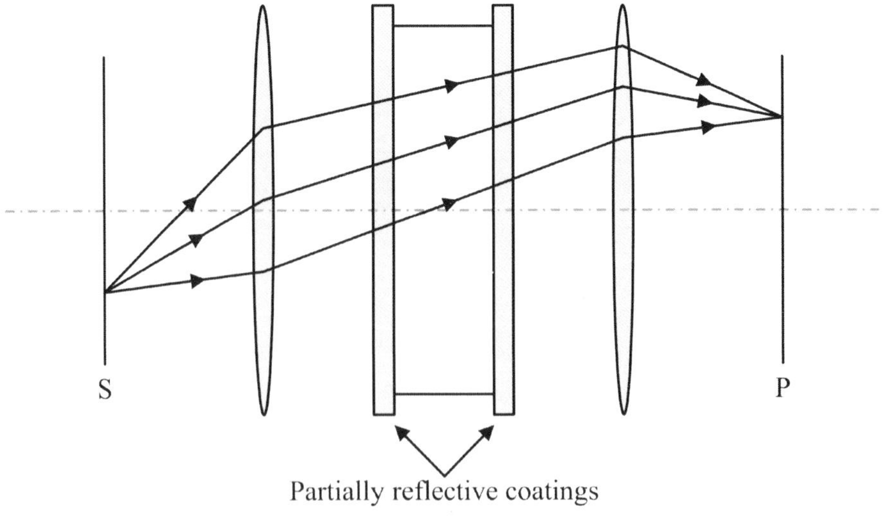

Figure 4.24. A simple Fabry-Pérot interferometer.

MULTILAYER THIN FILMS

Optical surfaces characterized by an arbitrary reflectance and transmittance coefficients can be produced by means of thin film coatings. These films are usually deposited onto a glass or metal substrates. It is well known that the antireflection coatings are deposited onto camera lenses and other optical instruments components. There are also other applications of thin films such as heat-reflecting and heat-transmitting mirrors ("hot" and "cold" mirrors), one-way mirrors, optical filters, etc.

First of all, we will consider the case of a single thin layer of dielectric material of index of refraction n and thickness t deposited on a semi-infinite substrate with the refractive index n_S as it is presented in figure 4.25. For the sake of simplicity we will consider an incident wave which propagates perpendicularly to the thin film. The modification for an arbitrary incident wave direction can be easily made. The electric field vector of the incident beam is $\mathbf{E}_I$ where the back reflected electric field vector is $\mathbf{E}_R$, and the transmitted electric field vector is equal to $\mathbf{E}_T$. The electric field vectors inside the thin film are $\mathbf{E}_T'$ and $\mathbf{E}_R'$, for the forward and backward propagating waves, respectively.

According to the boundary conditions, which require that the electric and magnetic fields to be continuous at each interface, for the first interface we have:

$$E_I + E_R = E_T' + E_R' , \qquad (4.92)$$

for electric fields amplitudes and:

$$H_I - H_R = H_T' - H_R' , \qquad (4.93)$$

for magnetic fields amplitudes, or equivalently:

$$E_I - E_R = n(E_T' - E_R') . \qquad (4.94)$$

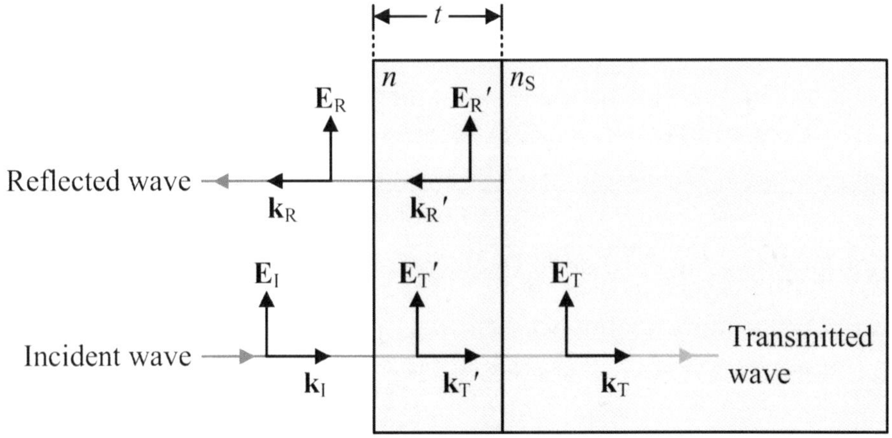

Figure 4.25. Wave vectors and their associated electric fields for the case of normal incidence on a single dielectric layer.

Consequently, for the second interface we have:

$$E_T' \exp(jkt) + E_R' \exp(-jkt) = E_T\,, \tag{4.95}$$

for electric fields amplitudes and:

$$H_T' \exp(jkt) - H_R' \exp(-jkt) = H_T\,, \tag{4.96}$$

for magnetic fields amplitudes, or equivalently:

$$n\left[E_T' \exp(jkt) - E_R' \exp(-jkt)\right] = n_S E_T\,. \tag{4.97}$$

By eliminating amplitudes inside the thin film se obtain:

$$\begin{aligned} 1 + \frac{E_R}{E_I} &= \frac{E_T}{E_I}\left[\cos(k\ ) - j\frac{n_S}{n}\sin(k\ )\right] \\ 1 - \frac{E_R}{E_I} &= \frac{E_T}{E_I}\left[n_S \cos(k\ ) - jn\ \sin(k\ )\right] \end{aligned}\,, \tag{4.98}$$

or, in matrix form:

$$\begin{bmatrix} 1 \\ 1 \end{bmatrix} + \begin{bmatrix} 1 \\ -1 \end{bmatrix}\frac{E_R}{E_I} = \begin{bmatrix} \cos(k\ ) & -j\dfrac{\sin(k\ )}{n} \\ -jn\ \sin(k\ ) & \cos(k\ ) \end{bmatrix}\begin{bmatrix} 1 \\ n_S \end{bmatrix}\frac{E_T}{E_I}\,, \tag{4.99}$$

which can be further reduced to:

$$\begin{bmatrix} 1 \\ 1 \end{bmatrix} + \begin{bmatrix} 1 \\ -1 \end{bmatrix} \rho = \mathbf{M} \begin{bmatrix} 1 \\ n_S \end{bmatrix} \tau \quad , \tag{4.100}$$

where we have introduced the reflection coefficient as:

$$\rho = \frac{E_R}{E_I} \quad , \tag{4.101}$$

and the transmission coefficient as:

$$\tau = \frac{E_T}{E_I} \quad . \tag{4.102}$$

The transfer matrix is also given by:

$$\mathbf{M} = \begin{bmatrix} \cos(k) & -j\frac{\sin(k)}{n} \\ -jn\sin(k) & \cos(k) \end{bmatrix} . \tag{4.103}$$

If we now assume that we have N layers of dielectric medium each having index of refraction n_i where $i = 1, 2, \ldots, N$, and thickness t_i we will derive the corresponding reflection and transmission coefficients in the same way as in the case of a single layer. Based on equation (4.100) one can show that the reflection and transmission coefficients are given by the following matrix equation:

$$\begin{bmatrix} 1 \\ 1 \end{bmatrix} + \begin{bmatrix} 1 \\ -1 \end{bmatrix} \rho = \mathbf{M}_1\mathbf{M}_2 \cdots \mathbf{M}_N \begin{bmatrix} 1 \\ n_S \end{bmatrix} \tau = \mathbf{M} \begin{bmatrix} 1 \\ n_S \end{bmatrix} \tau \quad , \tag{4.104}$$

where $\mathbf{M}_i$ corresponds to the transfer matrix of the i-th layer which have the same structure as given with equation (4.103) but with the corresponding index of refraction and thickness. For the overall transfer matrix $\mathbf{M}$ we can write the following:

$$\mathbf{M} = \begin{bmatrix} A & B \\ C & D \end{bmatrix} . \tag{4.105}$$

By slowing equations (4.104) and (4.105) in terms of the overall transmission matrix elements we have:

$$\rho = \frac{A - C + n_S(B - D)}{A + C + n_S(B + D)}$$

$$\tau = \frac{2}{A + C + n_S(B + D)} \quad (4.106)$$

The reflectance and the transmittance in terms of irradiances are gives as: $R = |\rho|^2$ and $T = |\tau|^2$. In the case of a single layer and according to equations (4.99) and (4.100) se have the following for the reflection coefficient:

$$\rho = \frac{n(1 - n_S)\cos(k\ ) - j(n_S - n^2)\sin(k\ )}{n(1 + n_S)\cos(k\ ) - j(n_S + n^2)\sin(k\ )} \quad (4.107)$$

If the thickness of the film is ¼ of the wavelength we have $kt = \pi/2$, so the reflectance in this case id given by:

$$R = |\rho|^2 = \frac{(n_S - n^2)^2}{(n_S + n^2)^2}, \quad (4.108)$$

which has the zero value, i.e. there is no reflection (antireflective coating) if it is fulfilled:

$$n = \sqrt{n_S}, \quad (4.109)$$

Magnesium fluoride is usually used for coating the lenses. As the lens has been built typically from a glass material having index of refraction of about $n_S \approx 1.5$ the requirement given by equation (4.109) isn't fulfilled as the magnesium fluoride has the index of refraction of $n = 1.35$. Nevertheless, with a quarter wave thickness of the magnesium fluoride coating the reflectance is reduced to about 1%. If a two layers coating has been used one can obtain zero reflectance for particular wavelength for a given coating materials. More layers provide more extensive possibilities. Therefore, by using three suitably chosen layers one can reduced the reflectance below 0.25% over almost entire visible spectrum.

In order to obtain a high reflectance value in a multilayer stack, a stack of alternating layers of high n_H and low n_L index of refraction has been built where each layer has the thickness equal to a quarter of wavelength, as shown in figure 4.26. The transfer matrices are all of the same form where the product of two adjacent is given by:

$$\mathbf{m} = \begin{bmatrix} 0 & -j\dfrac{1}{n_H} \\ -jn_L & 0 \end{bmatrix} \begin{bmatrix} 0 & -j\dfrac{1}{n_H} \\ -jn_L & 0 \end{bmatrix} = \begin{bmatrix} -\dfrac{n_H}{n_L} & 0 \\ 0 & -\dfrac{n_L}{n_H} \end{bmatrix}. \quad (4.110)$$

If the stack consists of 2*N* layers then the overall transfer matrix of the complete multilayer film is given by:

$$\mathbf{M} = \mathbf{m}^N = \begin{bmatrix} \left(-\dfrac{n_H}{n_L}\right)^N & 0 \\ 0 & \left(-\dfrac{n_L}{n_H}\right)^N \end{bmatrix}. \quad (4.111)$$

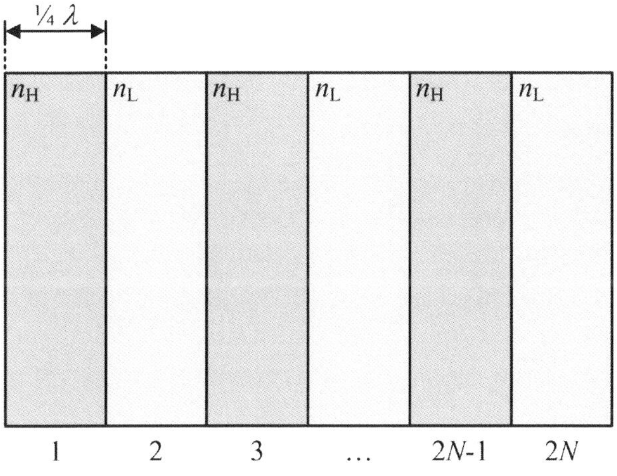

Figure 4.26. Multilayer stack for producing high reflectance.

According to equation (4.106) the reflectance of a multilayer film is given by:

$$R = |\rho|^2 = \left[\frac{\left(-\dfrac{n_H}{n_L}\right)^N - \left(-\dfrac{n_L}{n_H}\right)^N}{\left(-\dfrac{n_H}{n_L}\right)^N + \left(-\dfrac{n_L}{n_H}\right)^N}\right]^2 = \left[\frac{\left(\dfrac{n_H}{n_L}\right)^{2N} - 1}{\left(\dfrac{n_H}{n_L}\right)^{2N} + 1}\right]^2. \quad (4.112)$$

One can notice from the last equation that the reflectance approach unity for a large *N*. For example, for *N*=4 (eight=layers stack) of zinc sulfide (n_H = 2.3) and magnesium fluoride (n_L = 1.35) we have the reflectance of about R = 97% which is higher than the reflectance of a pure silver layer in the visible spectrum. Further, in the case of a 30-layers stack the reflectance is better than 99.9%. The maximum reflectance is obviously achieved for a single wavelength but it is possible to broad the range by combining different layers thicknesses.

INTERFEROMETRIC SENSORS

Being highly sensitive, interferometers can be used to measure different physical quantities with an ultimate precision. Typically, interferometers perform vibration amplitude measurement, flow velocity as well as flow velocity distribution measurement, rotation angles as well rotation velocity measurement. Moreover, today's state-of the art measurement such as gravitational wave detectors are interferometer based.

The Rayleigh Refractometer

The Rayleigh refractometer is based on the Young's double slit interferometer, although any other two-beam interferometer could be used. In order to measure the refractive index of materials having index of refraction close to unity such as gases, Rayleigh put in one of the arms, named as the sensing arm, the tube T_1 field with the gas which index of refraction is to be measured and in the other arm, named as the reference arm, the second tube T_2 filed with the gas with the known index of refraction or simply evacuated to form a vacuum, as presented in figure 4.27. The tubes T_1 and T_2 are in the separate light paths from the slits S_1 and S_2, illuminated coherently from a single source.

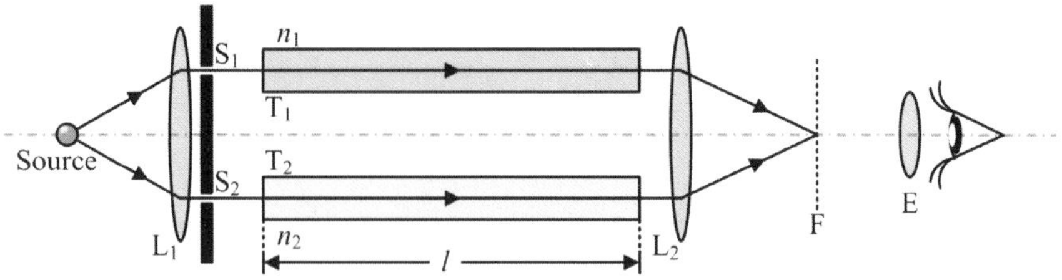

Figure 4.27. The Rayleigh refractometer.

If the gas pressure is changed in one of the tubes, the fringe system, viewed by an eyepiece E at the focus of a long-focus lens L_2, moves across the field of view. When moving, a fringes count N provides a direct measurement of the change in optical path through the tube caused by the change in refractive index δn as the amount of gas changes. If the tubes length is length l and wavelength of the light of the used source in vacuum λ_0 we have the following relation:

$$N = \frac{l\ddot{a}n}{\lambda_0} . \qquad (4.113)$$

In the case of a dilute gas the refractive index n differs from unity by an amount proportional to density, so that for a fixed temperature n-1 is proportional to pressure. In order to measure the gas pressure one can obtain the refractive index from a single measurement, which then can be used to calculate the value for any other pressure by a simple proportion.

Laser-Doppler Interferometry

The flow velocity can be easily measured by the laser-Doppler interferometry. This technique uses the fact that light scattered by a moving particle has its frequency shifted according to the Doppler effect. The frequency shift can be measured by the beats produced between the scattered light and a reference beam. A typical measurement configuration is based on two intersecting laser beams, making mutual angle θ, which illuminate the test field where the flow takes a part, as shown in figure 4.28. Light which is scattered by a particle passing through the beam overlapping region is focused on a photodetector. Assuming that particles have the same velocity v in the plane of the beams, at right angles to the direction of observation, the frequency of the beat signal is given by:

$$\Delta\nu = \frac{2v\sin\frac{\theta}{2}}{\lambda} . \qquad (4.114)$$

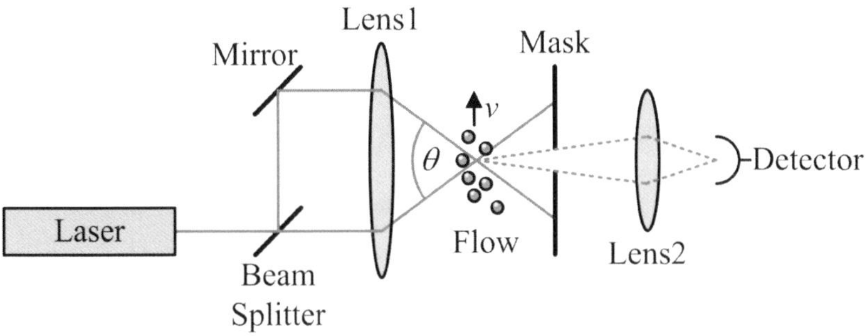

Figure 4.28. Laser-Doppler interferometer measures flow velocity.

In order to resolve the flow direction one can introduce a small frequency shift between two intersecting beams by a pair of acousto-optic modulators operated at slightly different frequencies. Simultaneous measurements of the velocity components along two orthogonal directions can be arranged by using two orthogonal illuminating beams. In order to avoid the beams interaction lasers with different wavelengths can be employed.

Vibration Amplitudes Measurements

Measurements of very small vibration amplitudes can be performed wit the help of interferometry. Typically, one of the beams in an interferometer is back reflected from a mirror attached to the object, which vibration amplitude is to be measured. Due to the Doppler effect there is a frequency shift of the back reflected beam. In order to measure the frequency shift the reflected beam is made to interfere with a reference beam with a known offset frequency. The photodetector time-varying output signal consists of a component at the offset frequency, i.e. the carrier and two sidebands. The vibration amplitude A can then be obtained from the following relation:

$$\frac{2\pi}{\lambda}A = \frac{I_{SB}}{I_0}, \tag{4.115}$$

where I_0 is the power at the offset frequency, and I_{SB} is the power in each of the sidebands.

Michelson's Stellar Interferometer

The principle of Michelson's stellar interferometer is shown in figure 4.29., which is generally based on a Young's double slits interferometer. However, there are some modifications of the simple Young interferometric system in order to make an interferometer capable of producing fringes by the light of an observed star. Instead of two narrow slits in the Young interferometer Michelson's stellar interferometer uses two large plane mirrors M_1 and M_2 positioned at a distance *D* from each other. Each mirror reflect light toward two mirrors M_3 and M_4, which further reflect two parallel beams of light into a telescope objective. Both beams form images of the star in the focal plane of the lens where typically a CCD camera can be placed. As a result fringes can be seen. Two apertures, positioned in front of the telescope objective, S_1 and S_2, limit the size of the beams, acting like Young's double slits interferometer with a spacing *d*.

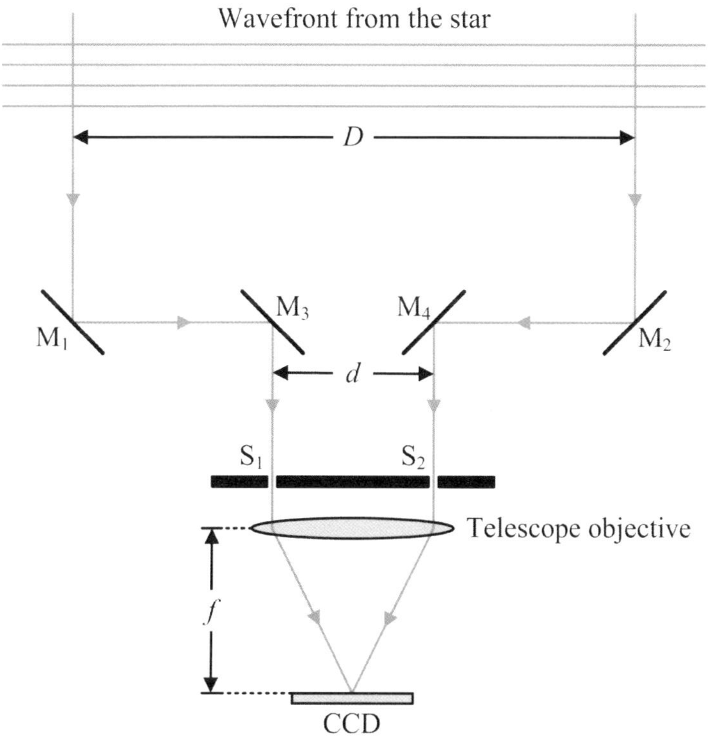

Figure 4.29. Michelson's stellar interferometer.

If the observed star has a very small angular diameter then the wavefronts falling onto mirrors M_1 and M_2 are coherent thus providing a coherent illumination of the slits S_1 and S_2. In the telescope objective focal plane, where the camera has been positioned, the interference pattern of the slits is observed, which consists of the fringes with a spacing $\lambda f/d$, where f is the focal length of the telescope.

In order measure the angular diameter of the star we will consider star as a light source which brightness distribution depends on the incoming angle ψ. Let this be $B(\psi)$ and let $B(\psi)$ be symmetrical about the center of the source. In this case each elementary strip of the source $d\psi$ wide and at a small angle ψ from the center of the source, forms a fringe pattern with the intensity proportional to $B(\psi)$ and displaced by angle ψ from center of the fringe system. Therefore, the fringe intensity maximum is given by:

$$I_{max} = C \int_{-\psi_0/2}^{\psi_0/2} B(\psi)\cos^2\left(\frac{\pi D}{\lambda}\psi\right)d\psi \quad , \qquad (4.116)$$

where C is a constant and ψ_0 is the star angular diameter. Similarly, fringe intensity minimum is given by:

$$I_{min} = C \int_{-\psi_0/2}^{\psi_0/2} B(\psi)\sin^2\left(\frac{\pi D}{\lambda}\psi\right)d\psi \quad . \qquad (4.117)$$

The fringe visibility of the observed inteferometric pattern is given by:

$$V = \frac{I_{max} - I_{min}}{I_{max} + I_{min}} = \frac{\int_{-\psi_0/2}^{\psi_0/2} B(\psi)\cos\left(\frac{2\pi D}{\lambda}\psi\right)d\psi}{\int_{-\psi_0/2}^{\psi_0/2} B(\psi)d\psi} \quad . \qquad (4.118)$$

For a star with uniform brightness we have:

$$V = \frac{\sin\left(\frac{\pi D}{\lambda}\psi_0\right)}{\frac{\pi D}{\lambda}\psi_0} \quad . \qquad (4.119)$$

Equation (4.119) represents the variation of fringe visibility with ψ_0 at a fixed spacing D or alternatively if D varies the change of fringe visibility can be used in order to measure angular diameter for a given star.

REFERENCES

1. Antoine Labeyrie, Stephen G. Lipson, and Peter Nisenson. *An Introduction to Optical Stellar Interferometry*. Cambridge: Cambridge University Press, 2006.
2. P. Hariharan. *Basics of Interferometry (2nd Edition)*. Amsterdam: Elsevier Inc., 2007.
3. Valery V. Tuchin (ed). *Coherent-Domain Optical Methods: Biomedical Diagnostics, Environmental and Material Science*. New York: Kluwer Academic Publishers, 2004.
4. Thomas Kreis. *Handbook of Holographic Interferometry: Optical and Digital Methods*. Weinheim: WILEY-VCH Verlag GmbH & Co.KGaA, 2005.
5. Igor A. Razumovsky. *Interference-Optical Methods of Solid Mechanics*. Heidelberg: Springer-Verlag Berlin Heidelberg, 2011.
6. Daniel Malacara, Manuel Servín, and Zacarias Malacara. *Interferogram Analysis for Optical Testing (2nd Edition)*. Boca Raton: Taylor & Francis Group, LLC, 2005.
7. P. Hariharan. *Optical Interferometry (2nd Edition)*. Amsterdam: Elsevier Science (USA), 2003.
8. F. Graham Smith, Terry A. King, and Dan Wilkins. *Optics and Photonics: An Introduction (2nd Edition)*. Chichester: John Wiley & Sons Ltd, 2007.
9. Andreas Glindemann. *Principles of Stellar Interferometry*. Heidelber: Springer-Verlag Berlin Heidelberg, 2011.
10. Richichi, F. Delplancke, F. Paresce, and A. Chelli (Eds.). *The Power of Optical/IR Interferometry: Recent Scientific Results and 2nd Generation Instrumentation - Proceedings of the ESO Workshop held in Garching, Germany, 4- 2005 8 April*. Heidelberg: Springer-Verlag Berlin Heidelberg, 2008.

5

CHAPTER

DIFFRACTION OF LIGHT

When an opaque object is placed between the point like light source and the screen one can notice that the shadow of the objected departs from the perfect sharpness of the edges as predicted by the geometrical optics. By careful inspection of the shadow edges there is also a certain amount of light that enters the dark zone behind the illuminated object, as well as that there are dark fringes that appear in the illuminated zone. This smearing of the shadow edge is closely related to the phenomena which cannot be explained in the frame of geometrical optics named diffraction. The essential qualitative characteristics of the diffraction phenomena can be explained in the frame of Huygens' principle. This basic phenomenon behind this principle is that the propagation of the light wave can be predicted as by assuming that each point on the wave front acts as a source of a secondary wave that spreads out in all direction equally, i.e. it behaves like a point like source. Therefore, the envelope of all the secondary waves is the new wave front.

HUYGENS' PRINCIPLE

In order to quantitatively treat the Huygens' principle we shall approach to the diffraction phenomena by introducing a precise mathematical form known as the Fresnel-Kirchhoff formula. This formula enables us to treat various specific cases of diffraction of light by obstacles and apertures. According to the Green's theorem if U and V are arbitrary two scalar-point functions, which satisfy usual conditions of differentiability and integrality he following is valid:

$$\oint_S (V\nabla U - U\nabla V)\mathrm{d}\mathbf{S} = \iiint_v \left(V\nabla^2 U - U\nabla^2 V\right)\mathrm{d}v \quad , \qquad (5.1)$$

where the left-hand integral extends over any closed surface S and the right-hand integral includes the volume v limited by the closed surface S. In the particular case where U and V are wave functions they satisfy the regular wave equation, so:

$$\nabla^2 U = \frac{1}{u^2}\frac{\partial^2 U}{\partial t^2}$$

$$\nabla^2 V = \frac{1}{u^2}\frac{\partial^2 V}{\partial t^2} \quad , \qquad (5.2)$$

where in the case of their harmonic time dependence it is easy to show that the volume integral in Green's theorem is identically zero. Therefore, the theorem reduces to:

$$\oint_S (V\nabla U - U\nabla V)\mathrm{d}\mathbf{S} = 0 \quad , \qquad (5.3)$$

If we take V to be the wave function in the form:

$$V = V_0 \frac{\exp[j(k \ + \omega t)]}{r} \quad , \qquad (5.4)$$

i.e. the presented function represents spherical waves that converges to the point P as presented in figure 5.1.

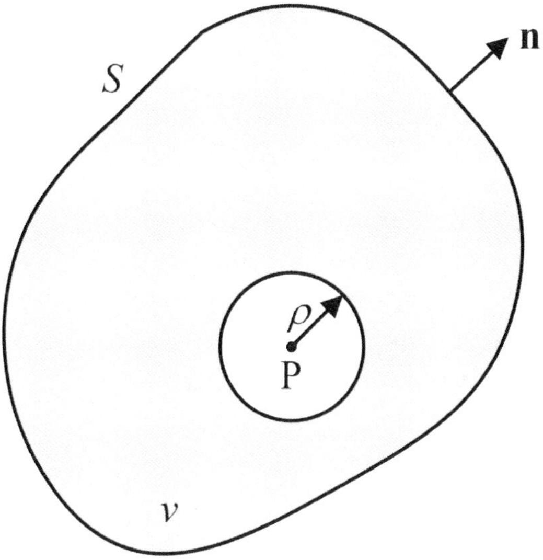

Figure 5.1. Surface of integration.

Now, let the volume v enclosed by the surface of integration S contains the point P. According to equation (5.4) one can notice that wave function V takes infinite value at the point P. In order to perform the integration we must exclude this point from the integration process. This can be accomplished by subtracting an integral over a small sphere of radius ρ and centered at the point P as shown in figure 5.1. Taking that for this small sphere $r = \rho$ and $\nabla = -(\partial/\partial r)\mathbf{n}$ is valid we have:

$$\oint_S \left\{ \frac{\exp(jkr)}{r} \nabla U - U\nabla\left[\frac{\exp(jkr)}{r}\right]\right\} \mathrm{d}\mathbf{S} - \oint_S \left\{ \frac{\exp(jkr)}{r}\frac{\partial U}{\partial r} - U\frac{\partial}{\partial r}\left[\frac{\exp(jkr)}{r}\right]\right\}_{r=\rho} \rho^2 \mathrm{d}\Omega = 0 \quad , \tag{5.5}$$

where dΩ is the elementary solid angle on the sphere centered at the point P and $\rho^2 \mathrm{d}\Omega$ is the corresponding elementary surface. If we further take the limit value for $\rho \to 0$, the value of the wave function U is constant within the small sphere and has the value U_P. In this case the left integral becomes:

$$\lim_{\rho\to 0} \oint_S \left\{ \frac{\exp(jkr)}{r}\frac{\partial U}{\partial r} - U\frac{\partial}{\partial r}\left[\frac{\exp(jkr)}{r}\right]\right\}_{r=\rho} \rho^2 \mathrm{d}\Omega = \oint_S U_P \mathrm{d}\Omega = 4\pi U_P \quad . \tag{5.6}$$

Equation (5.5) further develops to:

$$U_P = \frac{1}{4\pi} \oint_S \left\{ \frac{\exp(jkr)}{r} \nabla U - U\nabla\left[\frac{\exp(jkr)}{r}\right]\right\} \mathrm{d}\mathbf{S} \quad , \tag{5.7}$$

which is known as the Kirchhoff integral theorem. This theorem indicates that the value of any scalar wave function at any point inside an arbitrary closed surface is related to the value of the wave function at the surface. As the Kirchhoff integral theorem relates to the scalar fields it cannot be directly implemented to the electromagnetic filed, so the wave function U is usual known as the "optical disturbance". Nevertheless, the square of the absolute value of the wave function U may be related to the irradiance at a given point.

The Fresnel-Kirchhoff Formula

In order to mathematically teat the diffraction phenomena we will apply the Kirchhoff integral theorem. The diffraction phenomena are produced typically by an aperture of an arbitrary shape, which separates the light source from the observation point as it is presented in figure 5.2.

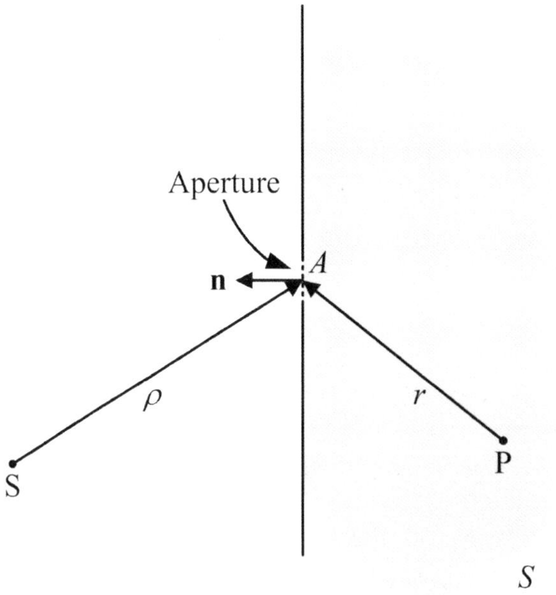

Figure 5.2. Geometry related to the Fresnel-Kirchhoff formula.

The general problem is to find the optical disturbance in the observation point P obtained from the light source S. When applying the Kirchhoff integral theorem we will consider the surface of integration S in such a way that it includes the observation point P as well as the aperture opening. The wave function from the source S at the aperture is given by:

$$U = U_0 \frac{\exp[j(k\rho - \omega t)]}{\rho} \quad , \tag{5.8}$$

where ρ is the position on the aperture relative to the source S position and represents the spherical monochromatic wave propagating outward from the source S. According to the Kirchhoff integral theorem we have:

$$U_P = \frac{U_0 \exp(-j\omega t)}{4\pi} \int_A \frac{\exp(jkr)}{r} \nabla \left[\frac{\exp(k\rho)}{\rho} \right] - \frac{\exp(k\rho)}{\rho} \nabla \left[\frac{\exp(jkr)}{r} \right] d\mathbf{A} \quad , \tag{5.9}$$

where the integration has been performed over the aperture opening A ($d\mathbf{A} = \mathbf{n}dA$) and where two basic simplifying assumption has been made:

- The wave function U and its gradient contribution is negligible except at the aperture openings and
- The wave function U and its gradient at the aperture has the same value as in the case of the partition absence.

The validity of these two assumptions is questionable. However, the experimental results justify them. The mathematical analysis gives the following relation for the integral functions:

$$\nabla \left[\frac{\exp(k\rho)}{\rho} \right] = \cos(\angle \mathbf{n}, \tilde{\mathbf{n}}) \frac{\partial}{\partial \rho} \left[\frac{\exp(k\rho)}{\rho} \right] = \cos(\angle \mathbf{n}, \tilde{\mathbf{n}}) \left[\frac{k \exp(k\rho)}{\rho} - \frac{\exp(k\rho)}{\rho^2} \right]$$

$$\nabla \left[\frac{\exp(jkr)}{r} \right] = \cos(\angle \mathbf{n}, \mathbf{r}) \frac{\partial}{\partial r} \left[\frac{\exp(jkr)}{r} \right] = \cos(\angle \mathbf{n}, \mathbf{r}) \left[\frac{k \exp(jkr)}{r} - \frac{\exp(jkr)}{r^2} \right] \quad , \tag{5.10}$$

where $\angle \mathbf{n}, \tilde{\mathbf{n}}$ and $\angle \mathbf{n}, \mathbf{r}$ denotes the angles between the corresponding position vectors ρ and **r**, respectively and the unity normal vector **n** on the aperture surface. In a typical case when the distances ρ and r are much larger then the wavelength of the light source, the second term in the parenthesis in equation (5.10), i.e. $\rho, r >> 1/k$ is fulfilled we obtain:

$$U_P = -k \frac{U_0 \exp(-j\omega t)}{4\pi} \int_A \frac{\exp[k(r+\rho)]}{r\rho} [\cos(\angle \mathbf{n}, \mathbf{r}) - \cos(\angle \mathbf{n}, \tilde{\mathbf{n}})] d\mathbf{A} \quad . \tag{5.11}$$

This equation is known as the Fresnel-Kirchhoff formula and actually represents the mathematical form of the Huygens' principle.

Babinet's Principle

Let a diffracting aperture A produces an optical disturbance U_P at a point of observation P. If the diffracting aperture is divided into two sub-apertures A_1 and A_2 that $A = A_1+A_2$ is fulfilled it is said that these two apertures are complementary as presented for example in figure 5.3. In this case from the Fresnel-Kirchhoff formula is obviously valid:

$$U_P = U_{P1} + U_{P2}\,, \qquad (5.12)$$

where U_{P1} is the optical disturbance at the point P which originates from the aperture A_1 only, and U_{P2} is the optical disturbance at the point P which originates from the aperture A_2 alone. This equation above is known as the Babinet's principle.

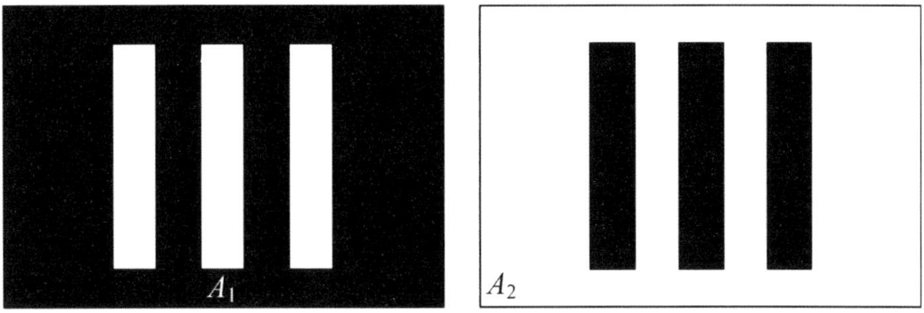

Figure 5.3. Complementary apertures.

One of the effects of the Babinet's principle is a paradox that in the case of diffraction limit, the radiation removed from the propagating beam due to a small particle is equal to twice the particle's cross section times the beam irradiance. The reason for this lies in the fact that the amount of radiation absorbed or reflected is the same as the amount diffracted.

FRAUNHOFER AND FRESNEL DIFFRACTION

Usually when considering diffraction phenomena it is customary to distinguish two general cases which are known as the Fraunhofer and Fresnel diffraction. Fraunhofer diffraction occurs in the case when both the incident and diffracted waves are effectively plane, as presented in figure 5.4.(a). Effectively this can be achieved when the distances from the source to the diffracting aperture and from diffracting aperture to the observation point are much larger that aperture dimensions so that their curvatures of the incident and diffracted waves can be neglected.

If the source and the receiving point are close enough to the diffraction aperture so that the curvature of the wave fronts are significant then we have the case of the Fresnel diffraction which is presented in figure 5.54. (b).

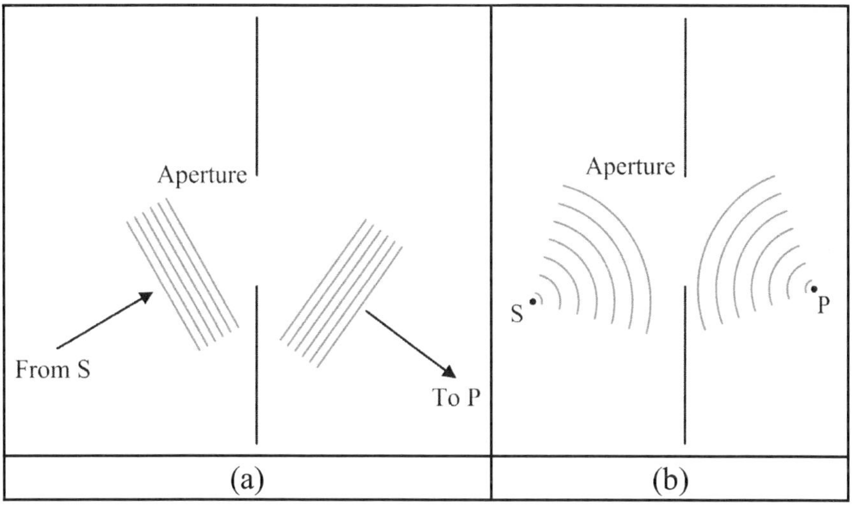

Figure 5.4. (a) Fraunhofer and (b) Fresnel diffraction.

One can see that there is no sharp distinction between the Fraunhofer and Fresnel diffraction. However, there is a quantitative measure in order to distinguish these two cases. Therefore, we will observe the diffraction geometry on the aperture opening as it is presented in figure 5.5. The observation point P is located at the distance p from the plane of the diffracting aperture whereas the source S is located at a distance s from this plane. The normal projection of the point source S to the plane is positioned at the distance x from the aperture edge, whereas the normal projection of the receiving point P to the plane is positioned at the distance y from the aperture edge. The size of the aperture openings is D.

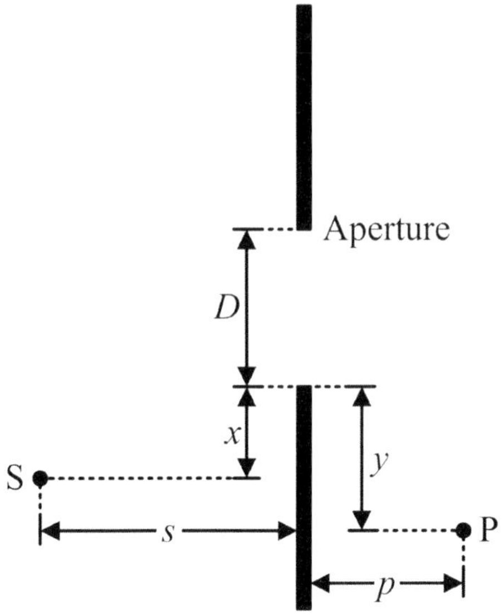

Figure 5.5. Geometry difference between the Fraunhofer and Fresnel diffraction.

According to the geometry presented in figure 5.5. the variation of the $r+\rho$ from one edge of aperture to the other is given by:

$$\Delta l = (r_1 + \rho_1) - (r_2 + \rho_2)$$

$$\Delta l = \left[\sqrt{s^2 + (x+D)^2} + \sqrt{p^2 + (y+D)^2}\right] - \left(\sqrt{s^2 + x^2} + \sqrt{p^2 + y^2}\right)$$

$$\Delta l \approx \left(\frac{x}{s} + \frac{y}{p}\right) D + \frac{1}{2}\left(\frac{1}{s} + \frac{1}{p}\right) D^2 + \ldots \qquad .(5.13)$$

The quadratic term in the expansion of the last equation essentially represents a measure of the curvature of the wavefront. If this quadratic term is negligible when compared with the wavelength of the light the wave is essentially plane over the diffracting aperture, so if the following condition is fulfilled:

$$\frac{1}{2}\left(\frac{1}{s} + \frac{1}{p}\right) D^2 < \lambda \quad , \qquad (5.14)$$

we have the Fraunhofer diffraction. In opposite case when the wavefront curvature takes a part in the diffraction we have the Fresnel type of diffraction. According to the Babinet's principle the same criterion is valid if instead of the aperture opening one have as opaque object. The criterion is also given by equation (5.14) where D represents the object dimension.

Fraunhofer Diffraction

In figure 5.6 is presented the typical setup for observing the Fraunhofer diffraction. The aperture is illuminated coherently by the monochromatic point light source S and a collimating lens. The plane wave of the collimated beam is diffracted at the aperture opening and further focused by the focusing length positioned behind the aperture. The Fraunhofer condition has been strictly fulfilled as the collimating lens with the point source in its focus produces beam consisted of parallel rays, i.e. plane wave is formed at the collimating lens output plane. In order to apply the Fresnel-Kirchhoff formula to simply calculate the diffraction patterns the following assumption will be taken into account:

- Spread angle of the diffracted rays is small enough so that the factor $[\cos(\angle\mathbf{n},\mathbf{r})-\cos(\angle\mathbf{n},\tilde{\mathbf{n}})]$ in equation (5.11) varies just slightly that we can take it outside the integral as well as
- The factor $\exp(jk\rho)/\rho$ also varies just slightly and also can be taken outside the integral, and
- The variation of the factor $\exp(jkr)/r$ originates mostly from the exponential part, so the factor $1/r$ can be omitted and replaced by its mean value.

By taking all these simplifications the Fresnel-Kirchhoff formula reduces to the following form:

$$U_{\mathrm{P}} = C\int_A \exp(jkr)\mathrm{d}A \quad , \qquad (5.15)$$

where all the constant factors have been replaced by a constant C.

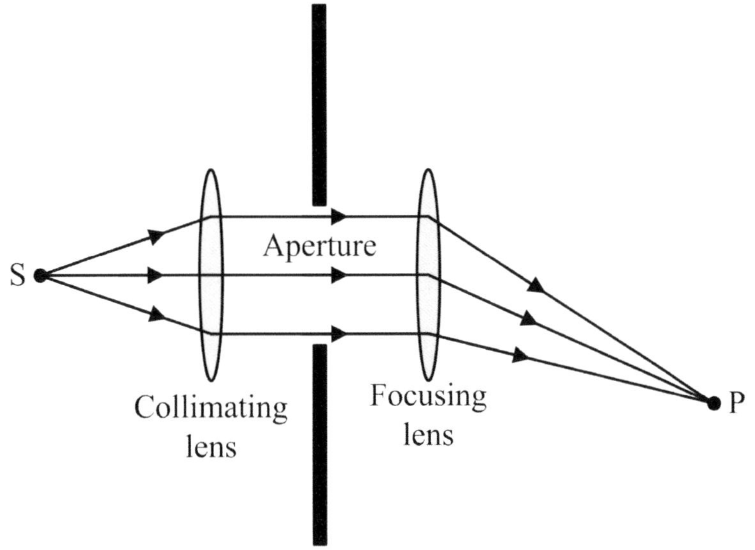

Figure 5.6. Fraunhofer diffraction arrangement.

In the case of diffraction by a single narrow slit, as presented in figure 5.7., we will consider a slit of the length L and width w. Therefore, the elementary area is $dA = Ldh$ and where the parameter r can be found as:

$$r = r_0 + h\sin\theta, \tag{5.16}$$

where r_0 is the parameter r for $h = 0$, and the angle θ is presented in figure 5.7. By taking this analysis into account, equation (5.15) becomes:

$$U_P = \boldsymbol{C}\ \exp(jkr_0)\int_{-w/2}^{w/2}\exp(jkh\sin\theta)dh$$

$$U_P = \boldsymbol{C}\ \exp(jkr_0)\frac{\sin\left(\frac{1}{2}kw\ \sin\theta\right)}{k\sin\theta} = K\frac{\sin\alpha}{\alpha}, \tag{5.17}$$

where $\alpha = kw\ \sin\theta/2$ and $K = CLw\exp(jkr_0)$. Consequently, as the light has been brought into the focus of the focusing lens, the corresponding irradiance distribution in the focal plane is then given by:

$$I_P = |U_P|^2 = I_0\left(\frac{\sin\alpha}{\alpha}\right)^2, \tag{5.18}$$

where $I_0 = |CLw|^2$ which represents the irradiance for θ = 0.

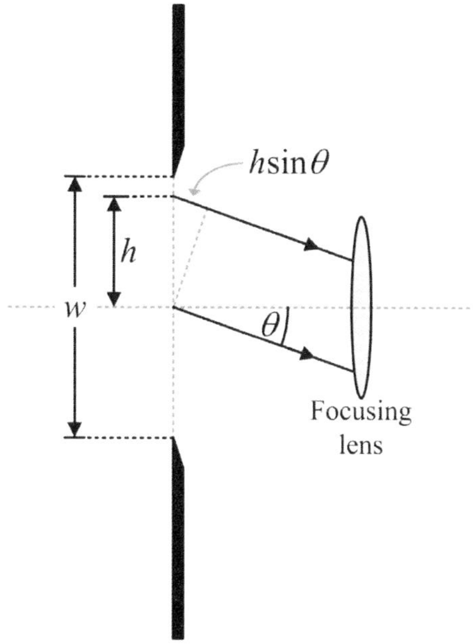

Figure 5.7. Fraunhoffer diffraction by a single slit.

Figure 5.8. represents the Fraunhofer diffraction pattern for a single slit. The maximum value occurs for θ = 0, where the zero values occur for α = kϖ, where k is an integer. As can be seen from figure 5.7., higher order maximums decrease rapidly and at the focal plane the diffraction pattern consists of a central bright part and at both sides one has alternating bright and dark parts. The first maximum occurs for α = ±π which corresponds to:

$$\sin\theta = \frac{2\pi}{kv} = \frac{\lambda}{w}, \quad (5.19)$$

which results in the fact that the angular width of the diffraction pattern varies inversely with the slit width where the amplitude of the central maximum is proportional to the slit area.

In the case of the rectangular aperture the mathematical formalism is very similar as in the case of a single slit. The only difference is that the integration must be performed in two dimensions as shown in figure 5.9. where the rectangular diffraction aperture has been presented. After the integration the obtained irradiance distribution in the focal plane is in the form of the product of two single slit distribution functions by:

$$I_P = I_B \left(\frac{\sin\alpha}{\alpha}\right)^2 \left(\frac{\sin\beta}{\beta}\right)^2, \qquad (5.20)$$

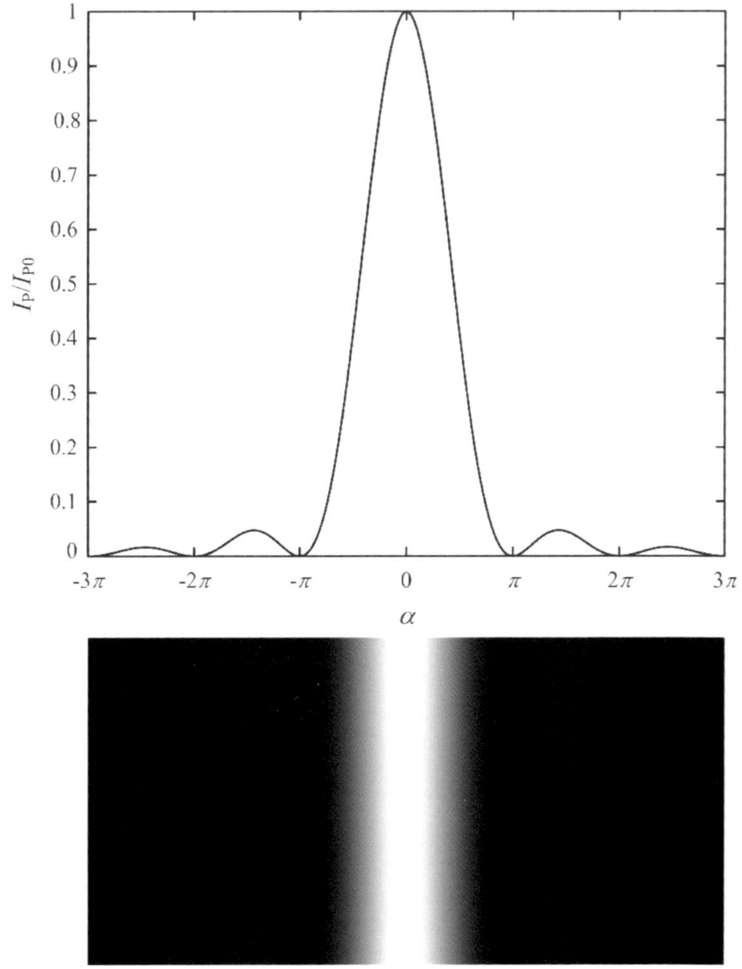

Figure 5.8. Fraunhoffer diffraction pattern on a single slit.

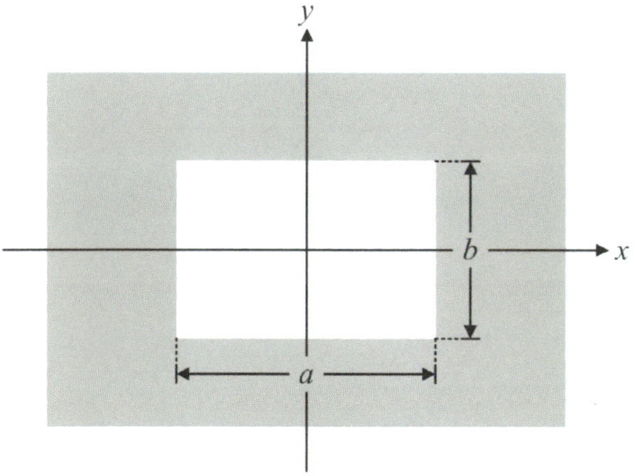

Figure 5.9. Rectangular diffracting aperture.

where $\alpha = ka \sin\xi/2$ and $\beta = ka \sin\psi/2$ and where the aperture dimensions are *a* and *b* where angles ξ and ψ defined the diffracted ray direction. The resulting diffraction pattern has been presented in figure 5.10 where the lines of zero irradiance are obtained for $\alpha = k\varpi$ and $\beta = m\varpi$ where *k* and *m* are integers.

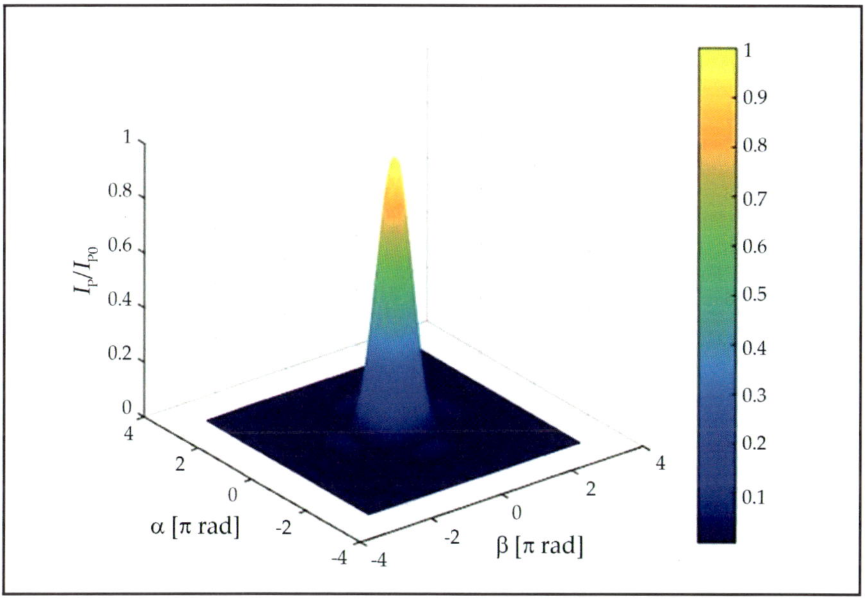

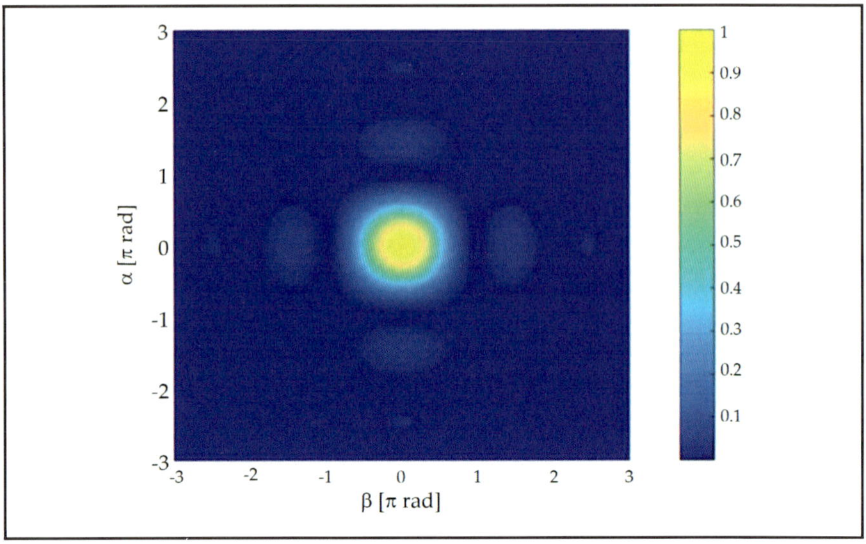

Figure 5.10. Fraunhoffer diffraction pattern on a rectangular slit.

In the case of a circular diffraction aperture in order to calculate the diffraction pattern we will observe the circular aperture opening as presented in figure 5.11. The aperture radius is R so the length of the marked strip will be given by:

$$l = 2\sqrt{R^2 - h^2}, \qquad (5.21)$$

where the strip width is dh.

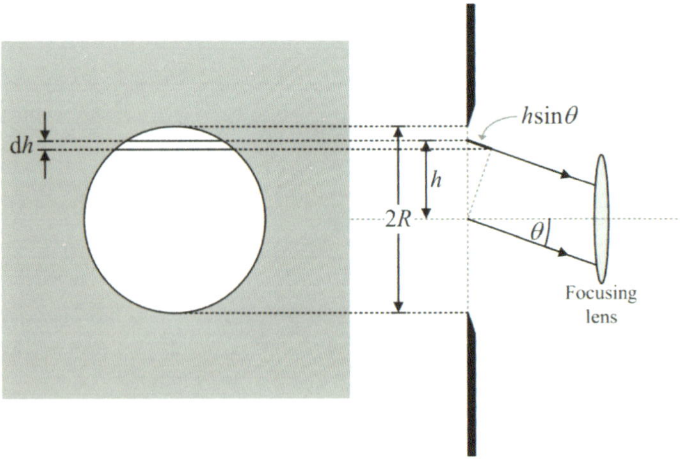

Figure 5.11. Circular diffraction aperture.

The diffraction pattern amplitude distribution is equal to:

$$U_P = C\exp(jkr_0)\int_{-R}^{R} 2\sqrt{R^2 - h^2}\exp(jkh\sin\theta)\mathrm{d}h \quad . \quad (5.22)$$

If we introduce the following substitutions $x = h/R$ and $y = kR\sin\theta$ the integral given by equation (5.22) becomes:

$$U_P \propto \int_{-1}^{1}\sqrt{1 - x^2}\exp(jxy)\mathrm{d}x \quad , \quad (5.23)$$

which represents the standard integral with the value equal to $\pi J_1(y)/y$ where $J_1(y)$ represents the Bessel function of the first kind and of the first order. As it is known $\lim_{y\to 0} J_1(y)/y = 1/2$ the irradiance distribution is given by:

$$I_P = I_{P0}\left[\frac{2J_1(y)}{y}\right]^2 \quad , \quad (5.24)$$

where $I_{P0} = |C\pi R|^2$ which is the irradiance for $\theta = 0$. The irradiance distribution for a circular opening is given in figure 5.12.

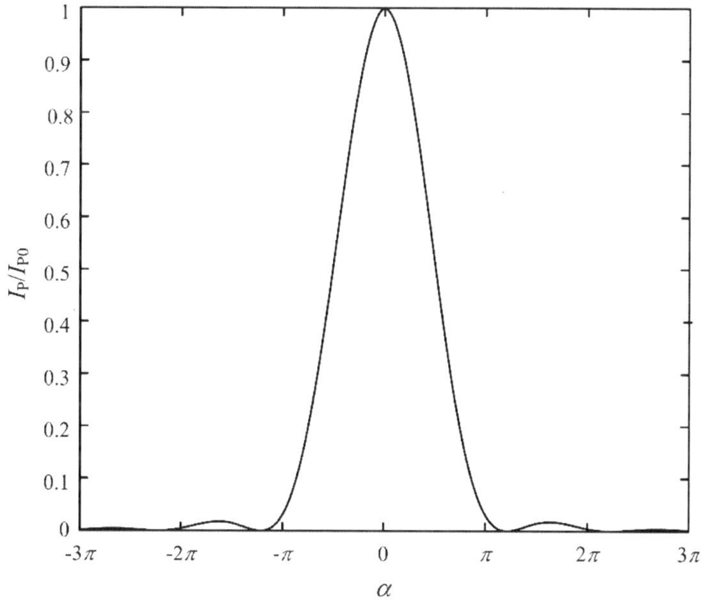

Figure 5.12. Fraunhoffer diffraction pattern on a circular slit.

The bright central area of the diffraction pattern, as presented in figure 5.12., is known as the Airy disk, which extends to the first dark ring whose size is defined by the first zero of the Bessel function $y = 3.832$. The angular radius of the first dark ring is given by:

$$\sin\theta \approx \theta \approx \frac{3.832}{kR} = \frac{1.2}{D}, \quad (5.25)$$

where D is the aperture diameter.

Let us further consider the diffraction aperture consisting of two parallel slits, as shown in figure 5.13., with the width w separated by the distance d. According to the Babinet's principle we have the following for the diffraction integral:

$$\int_A \exp(jky\sin\theta)\mathrm{d}y = \int_0^w \exp(jky\sin\theta)\mathrm{d}y + \int_d^{d+w} \exp(jky\sin\theta)\mathrm{d}y$$

$$= 2w\exp[j(\alpha+\beta)]\frac{\sin\alpha}{\alpha}\cos\beta \quad ,(5.26)$$

where $\alpha = (1/2)kw \sin\theta$ and $\beta = (1/2)kd \sin\theta$.

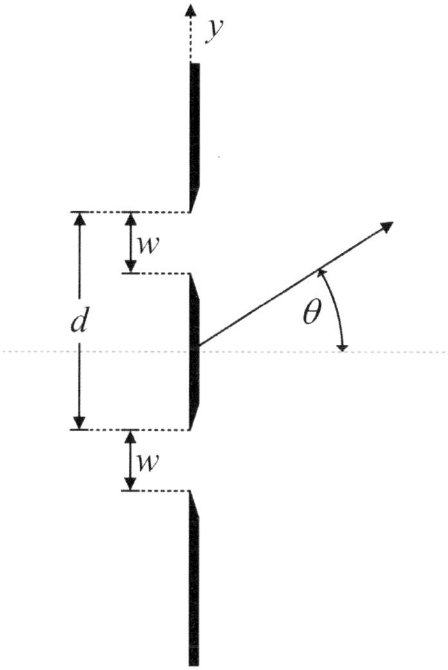

Figure 5.13. Double slit diffraction aperture.

The corresponding irradiance distribution of the diffraction parent of a double slit aperture is given by:

$$I_P = I_0 \left(\frac{\sin\alpha}{\alpha} \right)^2 \cos^2\beta \quad (5.27)$$

As in the case of a single slit the double slits diffraction pattern has also the factor $(\sin\alpha/\alpha)^2$ which represents at the same time the envelope of the diffraction fringes which are given by the term $\cos^2\beta$. The diffraction fringe pattern is presented in figure 5.14. in the case $d = w$. It is obvious that the bright fringes occur when $\beta = k\varpi$, where k is an integer ($\cos\beta = 1$). According to this analysis the angular separation between the fringes is approximately given by $\Delta\theta \approx \lambda/d$, which at the same time represents the result of analysis of the Young's double slit interferometer.

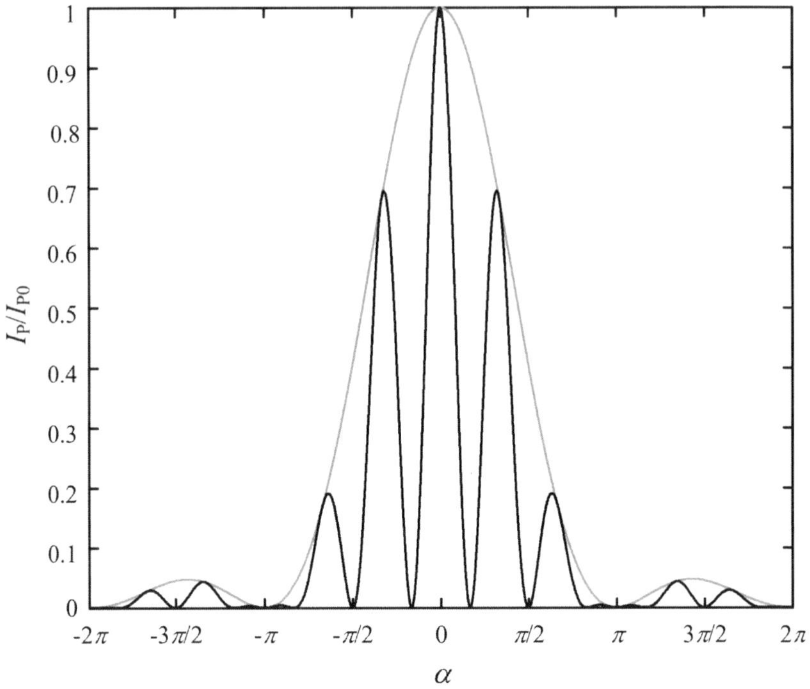

Figure 5.14. Fraunhoffer diffraction pattern on a double slit aperture.

If we now expand the previous analysis on the aperture that consists of *N* parallel slits (gratings) as presented in figure 5.15. with the width *w* and adjacent slits separation *d*. According to the Babinet's principle we have the following for the diffraction integral:

$$\int_A \exp(jky\sin\theta)dy = \sum_{i=1}^{N} \int_{(i-1)w}^{(i-1)w+d} \exp(jky\sin\theta)dy$$

$$= \frac{\exp(jkd\sin\theta)-1}{k\sin\theta}\sum_{i=1}^{N}\exp[k(i-1)d\sin\theta]$$

$$= \frac{\exp(jkd\sin\theta)-1}{k\sin\theta}\cdot\frac{1-\exp(jkNd\sin\theta)}{1-\exp(jkd\sin\theta)}$$

$$= w\exp\{j[\alpha+(N-1)\beta]\}\frac{\sin\alpha}{\alpha}\frac{\sin N\beta}{\sin\beta} \quad ,(5.28)$$

where $\alpha = (1/2)kw\ \sin\theta$ and $\beta = (1/2)kd\ \sin\theta$. The corresponding irradiance distribution of the diffraction parent of a double slit aperture is given by:

$$I_{\mathrm{P}} = I_{\mathrm{P0}} \left(\frac{\sin \alpha}{\alpha} \right)^2 \left(\frac{\sin N\beta}{N \sin \beta} \right)^2 , \qquad (5.29)$$

where I_{P0} is the irradiance for $\theta = 0$.

Figure 5.15. Multiple slits diffraction aperture.

As in the case of a double slit diffraction, we have also in the case of multiple slits the envelope of the diffraction pattern that performs an amplitude modulation of the diffraction fringes. The bright fringes occur when $\beta = k\varpi$, where k is an integer ($\cos\beta = 1$), or when it is fulfilled:

$$k\lambda = d \sin \theta , \qquad (5.29)$$

which represents the grating formula providing the relation between the wavelength and the diffraction angle and where k represents the order of diffraction.

The angular width of the fringe, i.e. the angular separation between the peak (maximum brightness) and the adjacent valley (minimum brightness), can be found when $N\beta = \pi$ is fulfilled, so we have:

$$\Delta\theta = \frac{\beta}{N\cos\theta}, \qquad (5.30)$$

where in the case of a very large N, $\Delta\theta$ becomes very small and the diffraction pattern consists of a series of sharp fringes, as presented in figure 5.16. for (a) $N = 3$, (b) $N = 5$, (c) $N = 10$, and (d) $N = 20$ and for a different diffraction order.

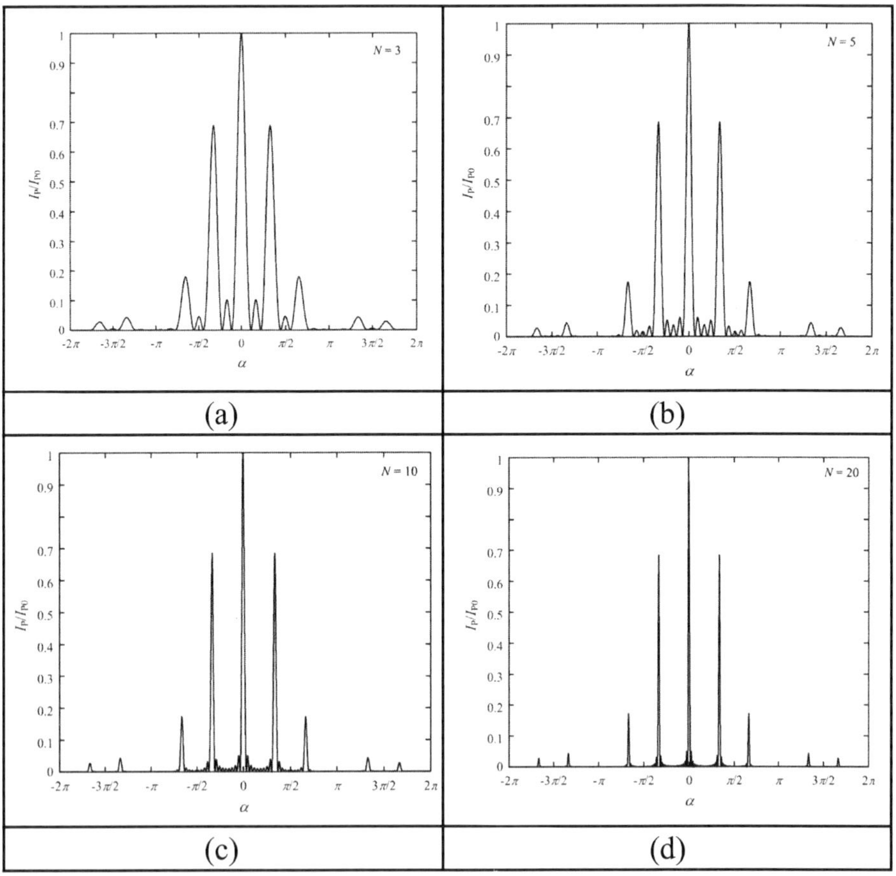

Figure 5.16. Fraunhofer diffraction patterns of a multiple slits aperture.

According to equation (5.29) for a given diffraction order k we have:

$$\Delta\theta = \frac{k\Delta\lambda}{d\cos\theta}, \qquad (5.31)$$

which represents the angular separation between two spectral lines with the wavelength shift of $\Delta\lambda$. By combining equations (5.30) and (5.31) we obtain the resolving power of a grating spectroscope according to the Rayleigh criterion as:

$$\boldsymbol{R} = \frac{\lambda}{\Delta\lambda} = kN \quad , \qquad (5.32)$$

or the resolving power is equal to the number of grooves multiplied by the diffraction order number.

FRESNEL DIFFRACTION

In the case of the Fraunhofer diffraction we have a plane wave incident on a diffraction aperture, i.e. the radius of curvature is infinite and doesn't affect the diffraction process. If the radius of curvature of the incident wave is comparable with the size of the diffraction aperture than the Fresnel diffraction occurs. As the radius of curvature must be incorporated in the mathematical model, the Fresnel diffraction is mathematically more comprehensive. However, the Fresnel diffraction is actually simpler to observe experimentally as one need only the light source and the aperture opening.

Further in the analysis we will observe a light from a point source S which illuminates the plane aperture where the light central beam is perpendicular to the plane intersecting it in the point C, as it is depicted in figure 5.17. The observation point P is positioned on the line that passes through the light source S and the central point C.

Let point A, which lies in the aperture plane, is positioned at the distance R from the central point C. In the presented geometrical arrangement, the distance $\mathrm{SAP} = r_1 + r_2$ can be expressed as:

$$\begin{aligned} r_1 + r_2 &= \sqrt{l_1^2 + R^2} + \sqrt{l_2^2 + R^2} \\ &= l_1 + l_2 + \frac{1}{2} R^2 \left(\frac{1}{l_1} + \frac{1}{l_2} \right) , \end{aligned} \qquad (5.33)$$

where $SC = l_1$ and $CP = l_2$. Now, if we observe regions bounded by the concentric circles defined in way that the parameter r_1+r_2 differs for the value equal to the half of the wavelength at it inner and outer region boundaries, these regions are called Fresnel zones. According to the last equation we have the following for the corresponding i-th Fresnel zone radius:

$$R_i = \sqrt{i\lambda L}, \quad (5.34)$$

where the parameter L is given by:

$$L = \frac{1}{\frac{1}{l_1} + \frac{1}{l_2}}, \quad (5.35)$$

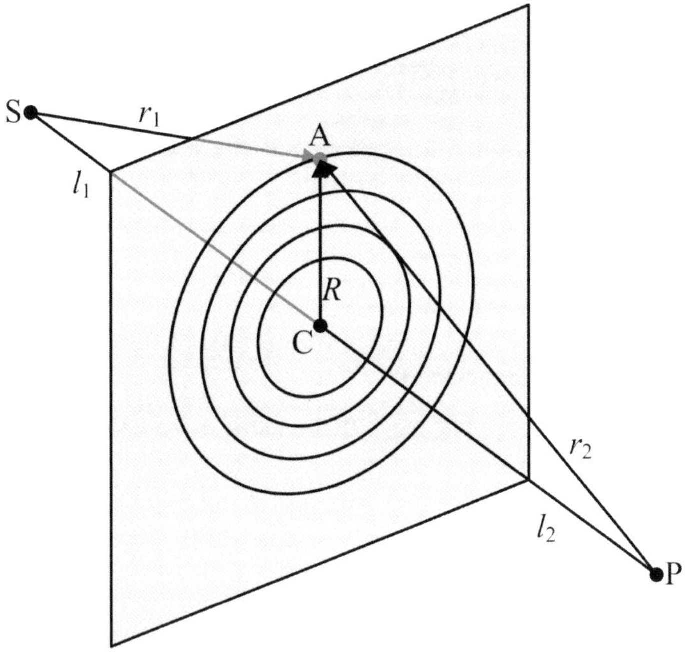

Figure 5.17. Fresnel zones in a plane aperture.

Further, from equation (5.34), if R_{i+1} and R_i are the outer and inner radius, respectively, of the i-th Fresnel zone then its area is given as $\pi R_{i+1}^2 - \pi R_i^2 = \pi R_1^2$, i.e. it is constant for any given i.

The optical disturbance at a given point of observation P can be found by finding the contribution of each Fresnel zone and summing them at the same time taking into account that the mean phase change between adjacent Fresnel zones is 180°. Therefore, the disturbance amplitude is given by:

$$|U_P| = \sum_i (-1)^{i-1} |U_i| \quad , \tag{5.36}$$

where U_i is the contribution of the i-th Fresnel zone. In the case when instead of an aperture we have a circular opaque object positioned symmetrically with regard to the line connecting the source S and the observation point P the disturbance at the point P is given by:

$$|U_P| = \sum_{i=k}^{+\infty} (-1)^{i-1} |U_i| \quad , \tag{5.37}$$

where the k-th zone is the first one that contributes the overall disturbance amplitude. According to the Fresnel-Kirchhoff formula, given by equation (5.11), the corresponding Fresnel zone amplitudes $|U_i|$ slowly varies with the increase of the zone order, so equation (5.37) can be rearranged as:

$$\begin{aligned} |U_P| &= \frac{(-1)^{k-1}}{2}|U_k| + \sum_{i=k,k+3,\ldots} \left[\frac{(-1)^{i-1}}{2}|U_i| + (-1)^i |U_{i+1}| + \frac{(-1)^{i+1}}{2}|U_{i+2}| \right] \\ &= \frac{(-1)^{k-1}}{2}|U_k| + \sum_{i=k,k+3,\ldots} (-1)^{i-1} \left[\frac{1}{2}|U_i| - |U_{i+1}| + \frac{1}{2}|U_{i+2}| \right] \\ &= \frac{(-1)^{k-1}}{2}|U_k| + \sum_{i=k,k+3,\ldots} (-1)^{i-1} \left[\frac{|U_i| + |U_{i+2}|}{2} - |U_{i+1}| \right] \\ &\approx \frac{(-1)^{k-1}}{2}|U_k| \end{aligned} \tag{5.38}$$

Equation (5.38) shows that at the center of the shadow of a circular opaque obstacle (point P) a bright spot occurs. The existence of such a spot was firstly theoretically predicted by Poisson in 1818 and later experimentally observed by Arago and Fresnel.

If the aperture is arranged in the form as presented in figure 5.18., i.e. when each odd (or even) Fresnel zone is obstructed, equation (5.36) gives the following amplitude in the observation point P:

$$|U_P| = \sum_i (-1)^{2(i-1)} |U_{2i-1}| = \sum_i |U_{2i-1}| \quad , \tag{5.39}$$

in the case when odd zones are obstructed. The zone plate acts very much like a lens as the amplitude is much larger than in the case when there is no aperture. The equivalent focal length of such a lens is the parameter L given by equation (5.35) and is equal to:

$$L = \frac{R_1^2}{\lambda} \quad . \tag{5.40}$$

Fresnel diffraction by a rectangular aperture can be analyzed by using the Fresnel-Kirchhoff formula in the Cartesian coordinate system as presented in figure 5.19. According to the presented geometry we have $R^2 = x^2 + y^2$ and thus:

$$\begin{aligned} r_1 + r_2 &= \sqrt{l_1^2 + R^2} + \sqrt{l_2^2 + R^2} \\ &= l_1 + l_2 + \frac{x^2 + y^2}{2L} \end{aligned} \quad . \tag{5.41}$$

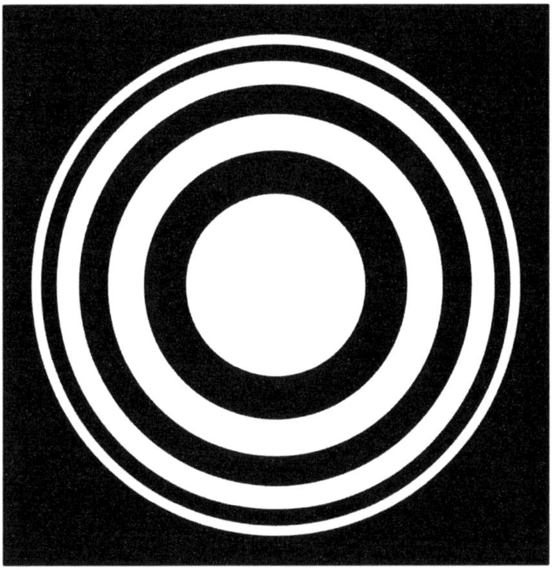

Figure 5.18. A zone plate.

By taking the same approximations as in the case of Fraunhofer diffraction the Fresnel-Kirchhoff formula becomes:

$$\begin{aligned} U_P &= C \int_{x=x_1}^{x_2} \int_{y=y_1}^{y_2} \exp\left(k \frac{x^2 + y^2}{2L} \right) \mathrm{d}x\mathrm{d}y \\ &= C \int_{x=x_1}^{x_2} \exp\left(k \frac{x^2}{2L} \right) \mathrm{d}x \int_{y=y_1}^{y_2} \exp\left(k \frac{y^2}{2L} \right) \mathrm{d}y \end{aligned} \quad , \tag{5.42}$$

where C is the constant that include all relevant factors.

By taking the following substitutions:

$$u = x\sqrt{\frac{k}{\pi L}}$$

$$v = y\sqrt{\frac{k}{\pi L}}, \qquad (5.43)$$

or equivalently:

$$u = x\sqrt{\frac{2}{\lambda L}}$$

$$v = y\sqrt{\frac{2}{\lambda L}}, \qquad (5.44)$$

equation (5.42) becomes:

$$U_P = K\int_{u=u_1}^{u_2} \exp\left(j\pi\frac{u^2}{2}\right)\mathrm{d}u \int_{v=v_1}^{v_2} \exp\left(j\pi\frac{v^3}{2}\right)\mathrm{d}v, \qquad (5.45)$$

where the constant K is given by $K = \lambda C\ /2$. The integrals given in equation (5.45) cannot be solved in the closed form but can be solved in terms of the following integral:

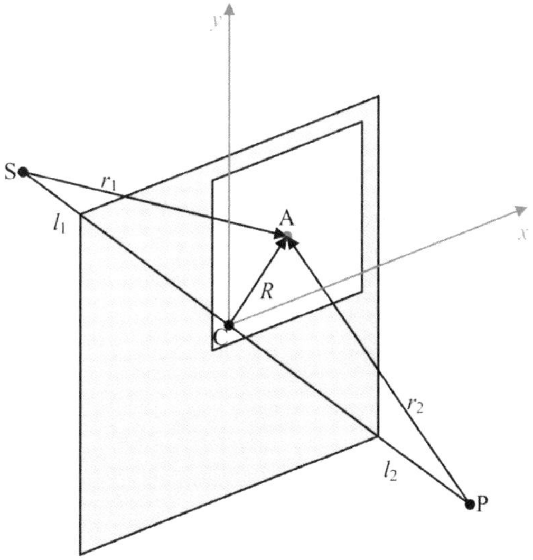

Figure 5.19. Diffraction on a rectangular aperture.

$$\int_0^s \exp\left(j\pi\frac{w^2}{2}\right)\mathrm{d}w = C(s) + jS(s) \quad , \qquad (5.46)$$

where the real and imaginary parts are given by:

$$C(s) = \int_0^s \cos\left(\frac{\pi}{2}w^2\right)\mathrm{d}w$$

$$S(s) = \int_0^s \sin\left(\frac{\pi}{2}w^2\right)\mathrm{d}w \quad , \qquad (5.47)$$

which are known as the Fresnel integrals. The graph which shows the dependence of the parameter C(s) versus the parameter S(s), known as the Cornu spiral, is presented in figure 5.20.

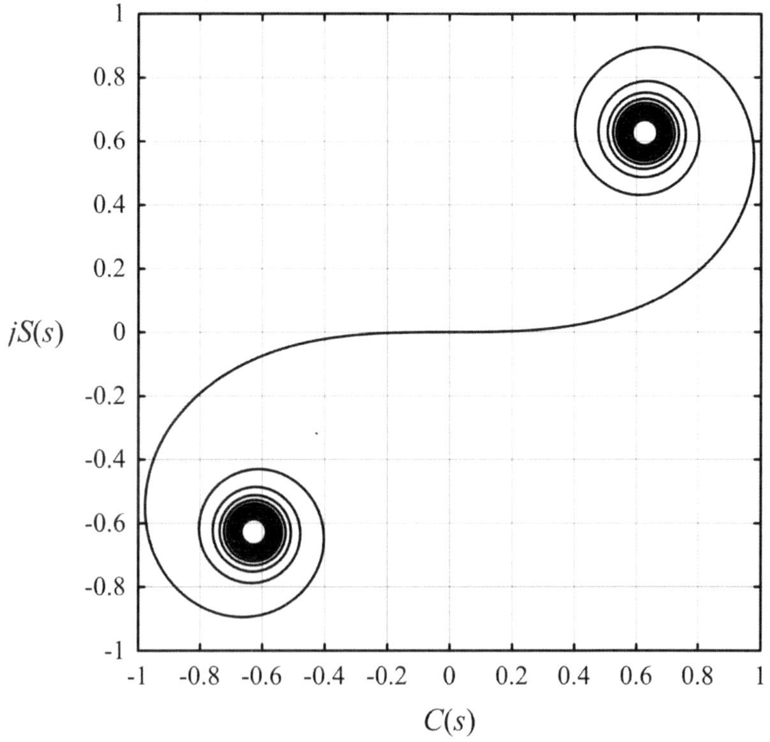

Figure 5.20. The Cornu spiral.

The values of the Fresnel integrals for the certain values of the parameter *s* are presented in table 5.1.

Table 5.1. Fresnel integrals

s	*C*(*s*)	*S*(*s*)
0.0	0.000	0.000
0.2	0.200	0.004
0.4	0.398	0.033
0.6	0.581	0.111
0.8	0.723	0.249
1.0	0.780	0.438
1.2	0.715	0.623
1.4	0.543	0.714
1.6	0.366	0.638
1.8	.0334	0.451
2.0	0.488	0.343
2.5	0.457	0.619
3.0	0.606	0.496
3.5	0.533	0.415
4.0	0.498	0.420
$+\infty$	0.500	0.500

REFERENCES

1. M. Françon. *Diffraction: Coherence in Optics*. Oxford: Pergamon Press, 1966.
2. John M. Cowley. *Diffraction Physics (3rd Revised Edition)*. Amsterdam: Elsevier Science B.V., 1995.
3. Okan K. Ersoy. Diffraction, Fourier Optics and Imaging. Hoboken: John Wiley & Sons, Inc., 2007.
4. V Minin and I V Minin. *Diffractional Optics of Millimetre Waves*. Bristol: IOP Publishing Ltd, 2004.
5. Swapan K. Saha. *Diffraction-Limited Imaging with Large and Moderate Telescopes*. New Jersey: World Scientific Publishing Co. Pte. Ltd., 2007.

6. Donald C. O'Shea, Thomas J. Suleski, Alan D. Kathman, and Dennis W. Prather. *Diffractive Optics: Design, Fabrication, and Test*. Bellingham: SPIE – The International Society for Optical Engineering, 2004.
7. Igor Minin and Oleg Minin. *Diffractive Optics and Nanophotonics: Resolution Below the Diffraction Limit*. Cham: Springer International Publishing AG Switzerland, 2016.
8. Pyotr Ya. Ufimtsev. *Fundamentals of the Physical Theory of Diffraction (2nd Edition)*. Hoboken: John Wiley & Sons, Inc., 2014.
9. Kostas Daniilidis and Reinhard Klette (Eds.). *Imaging Beyond the Pinhole Camera*. Dordrecht: Springer, 2006.
10. Eric J. Mittemeijer and Udo Welzel (Eds.). *Modern Diffraction Methods*. Weinheim: Wiley-VCH Verlag & Co. KGaA, 2013.
11. Virendra N. Mahajan. *Optical Imaging and Aberrations: Part II - Wave Diffraction Optics (2nd Edition)*. Bellingham: Society of Photo-Optical Instrumentation Engineers, 2011.
12. Elena Sokolova. *Optimization Techniques for Diffraction Spectrometers*. Bellingham: Society of Photo-Optical Instrumentation Engineers (SPIE), 2015.
13. Graeme L. James. *Geometrical Theory of Diffraction for Electromagnetic Waves*. Herts: Peter Peregrinus Ltd., 1976.
14. Fernando Moreno and Francisco González (Eds.). *Light Scattering from Microstructures*. Berlin: Springer-Verlag Berlin Heidelberg, 2000.

6
CHAPTER

HOLOGRAPHY

HOLOGRAPHIC IMAGING

Usually a hologram is captured by a two-dimensional photographic plate or a flat film. However, the hologram produces a three-dimensional image. Moreover, the hologram doesn't capture an image in a conventional way. In the case of a conventional imaging technique, such as photography, one records the radiant intensity distribution of the scene. The resulting image lost all the information about the optical paths to different parts of the scene. The idea behind the holography is to record both parameters: the phase and the amplitude of the light waves coming from an object. As all image capturing materials respond only to the irradiance one must to convert the phase information in some kind of the irradiance variations. In order to achieve this, a coherent illumination must be employed, as shown in figure 6.1., where the photographic film records the interference pattern produced by the reference beam and the light waves scattered by the object.

The irradiance any point of the captured interferometric pattern depends on the phase of the object wave so the recorded hologram contains information on the phase as well as the amplitude of the object wave. If the hologram is illuminated with the original reference wave, as presented figure 6.2, it reconstructs the original object wave.

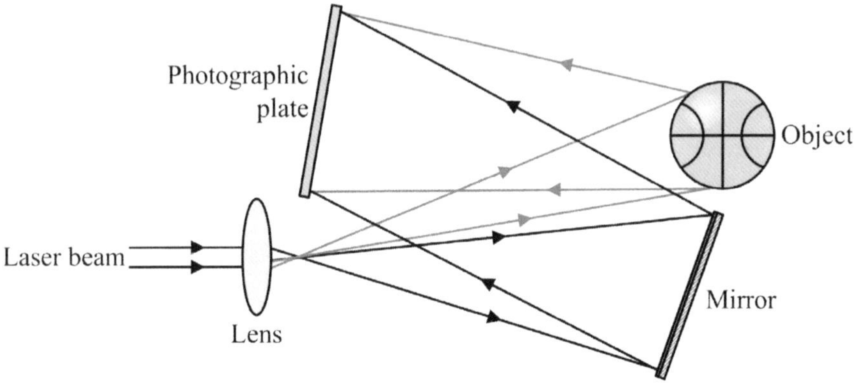

Figure 6.1. Recording of the hologram.

At the beginning, in order to simplify the discussion on the theory of holography, we will assume that the reference beam is collimated, i.e. it consists of plane waves, although this is not actually necessary in practice and as it is presented in figure 6.1. If we denote x and y to be the coordinates in the plane of the recording photographic plate and $U(x,y)$ the complex wave field of the object reflected wave front we have:

$$U(x,y) = a(x,y)\exp[j\varphi(x,y)], \qquad (6.1)$$

where $a(x,y)$ is the amplitude of the object wave field and $\varphi(x,y)$ is its phase.

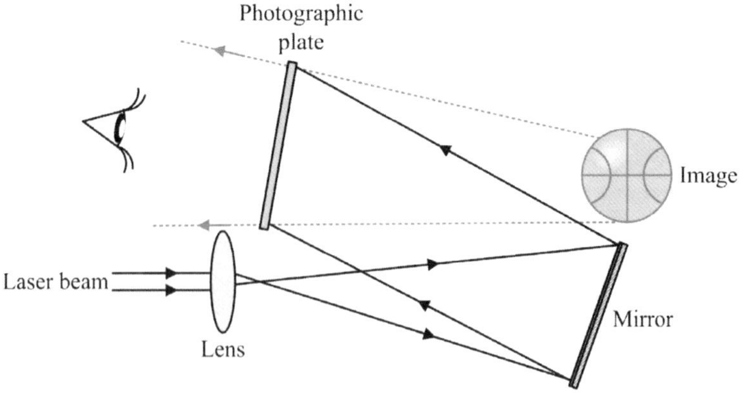

Figure 6.2. Image reconstruction.

Similarly, we will denote $U_0(x,y)$ to be the complex wave field of the reference beam in the photographic plate plane, for which is valid:

$$U_0(x,y) = a_0 \exp[j(\mu x + \nu y)], \quad (6.2)$$

where a_0 is the amplitude of the plane reference wave field and μ and ν are the spatial frequencies of the reference beam in the photographic plate plane which are given by:

$$\mu = k \sin\alpha$$
$$\nu = k \sin\beta, \quad (6.3)$$

where k is the wave number of the laser light and α and β are the direction angles of the reference beam. The corresponding irradiance that falls onto the photographic plate and which is also captured by the plate is given by:

$$\begin{aligned} I(x,y) &= |U(x,y) + U_0(x,y)|^2 \\ &= a_0^2 + a^2(x,y) + a_0 a(x,y)\exp\{j[\varphi(x,y) - (\mu x + \nu y)]\} + a_0 a(x,y)\exp\{-j[\varphi(x,y) - (\mu x + \nu y)]\} \\ &= a_0^2 + a^2(x,y) + 2a_0 a(x,y)\cos[\varphi(x,y) - (\mu x + \nu y)] \end{aligned},$$

$$(6.4)$$

which in fact represents an interferometric pattern. As can be seen from equation (6.4), the interference pattern contains the information in the form of amplitude and phase modulations of the spatial frequencies of the reference beam.

When the developed hologram is illuminated with a beam similar to the reference one, used for the hologram recording, the transmitted wave is proportional to the amplitude of the illuminating beam times the transmittance of the hologram, where the hologram transmittance is proportional to $I(x,y)$. Therefore, the transmuted wave amplitude is given by:

$$\begin{aligned} U_T(x,y) &\propto U_0(x,y) I(x,y) \\ &\propto a_0\left[a_0^2 + a^2(x,y)\right]\exp[j(\mu x + \nu y)] + a_0^2 a(x,y)\exp[j\varphi(x,y)] + a_0^2 a(x,y)\exp\{-j[\varphi(x,y) - 2(\mu x + \nu y)]\} \\ &\propto \left[a_0^2 + a^2(x,y)\right]U_0(x,y) + a_0^2 U(x,y) + a^2(x,y)\left[U_0^*(x,y)\right]^2 U^*(x,y) \end{aligned}.$$

$$(6.5)$$

where the star denote the complex conjugate operator. As a matter of fact the hologram acts in a way like diffraction grating where it produces the direct beam and two first order diffracted beams on either side of the direct beam. The direct beam is given with the term $\left[a_0^2 + a^2(x,y)\right]U_0(x,y)$

and one of the diffracted beams is given with the term $a_0^2 U(x,y)$. As this term is equal to the value of the complex wave field of the object reflected wave front multiplied with the constant a_0^2, this diffracted beam is the one that forms the virtual three-dimensional image. The term $a^2(x,y)U_0^{-2}(x,y)U^{-1}(x,y)$ represents the other diffracted beam, which forms the real image.

THE IN-LINE HOLOGRAM

The optical system aimed for recording an in-line hologram is presented in figure 6.3. The object, which contains small opaque elements on a clear background, is illuminated by a collimated beam of monochromatic light which is parallel to the main optical axis and perpendicular to the photographic plate. Due to such arrangement the light that impinges photographic plate consists of two components. The first one is the directly transmitted wave and represents the plane wave of the reference beam whose amplitude and phase do not vary across the photographic plate and which complex wave field is U_0. The second one is a weak scattered wave whose complex wave field at any point (x,y) on the photographic plate can be written as $U(x,y)$, where $|U_0| > |U(x,y)|$ is fulfilled. As the resultant complex wave field is obtained as the sum of the complex wave fields of the reference and scattered waves, the corresponding irradiance at the photographic plate is given by:

$$\begin{aligned} I(x,y) &= |U_0 + U(x,y)|^2 \\ &= |U_0|^2 + |U(x,y)|^2 + U_0 U(x,y) + U_0^* U^*(x,y), \end{aligned} \qquad (6.6)$$

where star denote the complex conjugate operator.

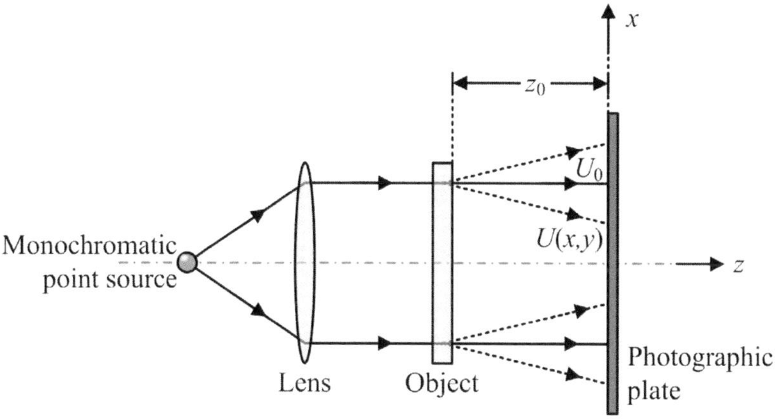

Figure 6.3. Optical system aimed for an in-line hologram recording.

The photographic plate reacts in a way that due to the irradiance that falls onto its surface forms a transparent zones whose transparency $t(x,y)$ depends on the irradiance in the following way:

$$t(x,y)=t_0+\alpha\, I(x,y), \tag{6.7}$$

where t_0 is a constant background transmittance, τ is the exposure time, and α is a parameter determined by the photographic plate material used and the processing conditions. Equations (6.6) and (6.7) give the amplitude transmittance of the hologram as:

$$t(x,y)=t_0+\alpha\left[|U_0|^2+|U(x,y)|^2+U_0U(x,y)+U_0^*U^*(x,y)\right]. \tag{6.8}$$

In order to reconstruct the image of an object, the hologram is illuminated as shown in figure 6.4. with the same collimated beam of monochromatic light as used to record the original hologram. The complex wave field at any given point within the illuminating beam is then given by:

$$\begin{aligned} U'(x,y) &= U_0 t(x,y) \\ &= U_0\left\{t_0+\alpha\left[|U_0|^2+|U(x,y)|^2+U_0U(x,y)+U_0^*U^*(x,y)\right]\right\} \\ &= U_0\left(t_0+\alpha\,|U_0|^2\right)+\alpha\,U_0|U(x,y)|^2+\alpha\,U_0^2U(x,y)+\alpha\,|U_0|^2U^*(x,y). \end{aligned} \tag{6.9}$$

The overall complex wave field consists of four terms as can be seen from equation (6.90). The first one $U_0\left(t_0+\alpha\,|U_0|^2\right)$ corresponds to a uniformly attenuated plane wave of the directly transmitted beam.

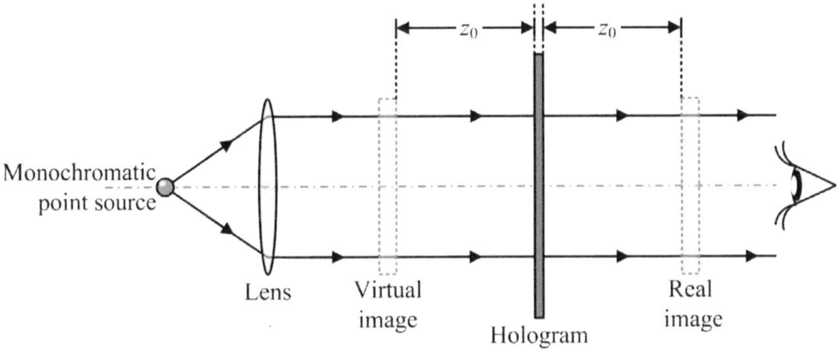

Figure 6.4. Optical system used for reconstructing the image with an in-line hologram.

The second one $\propto U_0|U(x,y)|^2$ can be omitted as it is extremely small ($|U(x,y)| < |U_0|$). The third one $\propto U_0^2U(x,y)$ is, except of a constant factor $\propto U_0^2$, identical to the complex amplitude of the scattered wave from the object and reconstructs an image of the object in its original position. This image is formed behind the hologram. Therefore, the reconstructed waves appear to diverge from it, so it represents a virtual image, as it is presented in figure 6.4. The fourth term $\propto |U_0|^2U^*(x,y)$ corresponds to a wave similar to the object wave but with the curvature opposite in sign. This wave converges to form a real image at the same distance in front of the hologram, as it is presented in figure 6.4.

The main drawback of the in-line hologram is that an observer see superimposed virtual and real image in the presence of a strong coherent background. Beside this drawback, there is also another that the object must have a high average transmittance in order to provide that the second term on the right-hand side of equation (6.9) to be negligible.

OFF-AXIS HOLOGRAM

The arrangement for recording of an off-axis hologram is presented in figure 6.5., where, in order to simplify the analysis, the same collimated reference beam with the uniform irradiance is used to illuminate the photographic plate and the object simultaneously. The complex wave field at a given point (*x*,*y*) on the photographic plate induced by the reference beam is given by:

$$U_R(x,y) = U_0 \exp(j2\pi\xi x), \qquad (6.10)$$

where $\xi = (\sin\theta)/\lambda$, as only the phase of the reference beam varies across the photographic plate.

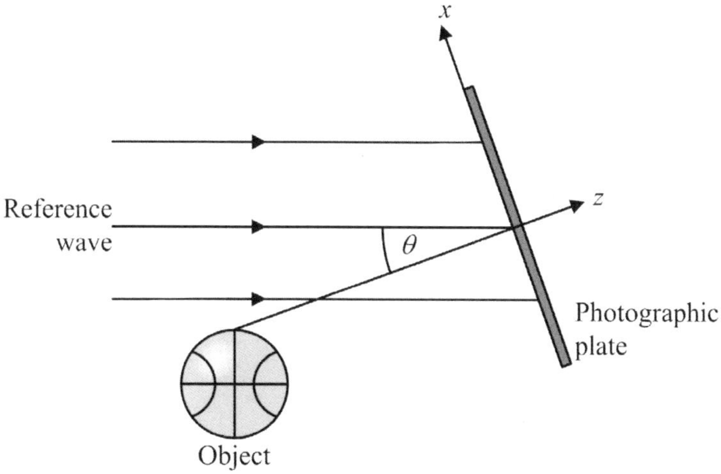

Figure 6.5. The recording of an off-axis hologram.

The complex wave field at a given point (*x*,*y*) on the photographic plate induced by the beam reflected from the object is given by:

$$U_O(x,y) = |U_O(x,y)| \exp[-j\varphi(x,y)], \qquad (6.11)$$

where both the amplitude and the phase of the object reflected complex wave field varies across the photographic plate. Therefore, the resultant irradiance is given by:

$$\begin{aligned} I(x,y) &= |U_R(x,y) + U_O(x,y)|^2 \\ &= |U_R(x,y)|^2 + |U_O(x,y)|^2 + \\ &+ U_0^*|U_O(x,y)|\exp[-j\varphi(x,y)]\exp(-j2\pi\xi x) + U_0|U_O(x,y)|\exp[j\varphi(x,y)]\exp(j2\pi\xi x) \\ &= |U_0|^2 + |U_O(x,y)|^2 + 2|U_0||U_O(x,y)|\cos[2\pi\xi x + \varphi(x,y)] \end{aligned} \qquad .(6.12)$$

Similarly, if we take that the amplitude transmittance of the developed photographic plate is linear function of impinging irradiance, then the transmittance of the hologram is given by:

$$t(x,y) = t_0 + \alpha\, |U_0|^2 + \alpha\, |U_O(x,y)|^2 + $$
$$+ \alpha\, U_0^*|U_O(x,y)|\exp[-j\varphi(x,y)]\exp(-j2\pi\xi x) + \alpha\, U_0|U_O(x,y)|\exp[j\varphi(x,y)]\exp(j2\pi\xi x). \quad (6.13)$$

If we illuminate the recorded hologram with the original reference beam, as shown in figure 6.6., the complex wave field of the transmitted wave can be written as:

$$U_T(x,y) = U_R(x,y)t(x,y)$$
$$= \left(t_0 + \alpha\, |U_0|^2\right)U_0\exp(j2\pi\xi x) + \alpha\, U_0|U_O(x,y)|^2\exp(j2\pi\xi x) +$$
$$+ \alpha\, |U_0|^2 U_O(x,y) + \alpha\, U_0^2 U_O^*(x,y)\exp(j4\pi\xi x) \quad . \quad (6.14)$$

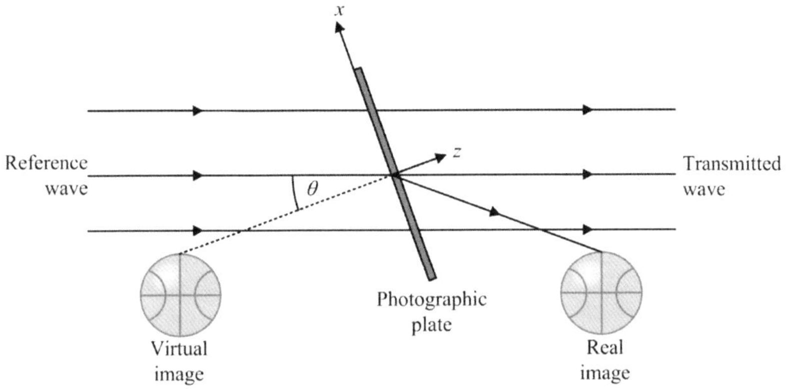

Figure 6.6. Image reconstruction of an off-axis hologram.

By inspecting the equation (6.14) one can notice that the first term on the right-hand side represents the directly transmitted beam, the second term generates a halo surrounding it, the third term is the one that corresponds to the original object wave but multiplied with the constant factor. This wave produces a virtual image of the object in its original position. The fourth term represents conjugate image which corresponds to a real image. The real and virtual images are located at the same distance from the hologram.

FOURIER HOLOGRAMS

One of the possible hologram recording configurations is presented in figure 6.7. where the interfering waved are the Fourier transforms of the of the complex weave field of the original object and of the reference wave. Usually the object must have limited thickness as well as transparency.

In order to record the Fourier hologram the object is placed in the front focal point of the lens and illuminated with a collimated and monochromatic light beam. The reference beam is obtained from a point source also located in the front focal plane of the lens. The Fourier hologram is recorded on a photographic plate which is placed in the lens back focal plane.

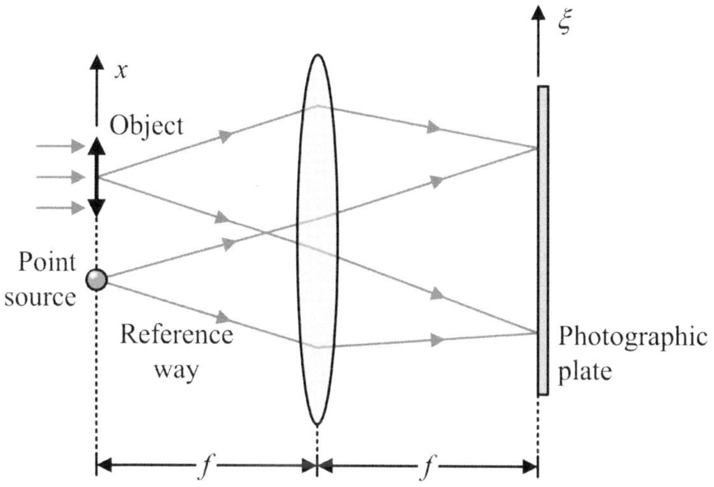

Figure 6.7. Fourier hologram recording optical system.

Let the complex wave field that leaves the object plane is $U(x,y)$ then its complex wave field at the photographic plate is given by:

$$u(\xi,\psi) = F\{U(x,y)\}, \qquad (6.15)$$

where $F\{\bullet\}$ represents the Fourier transform operator. The monochromatic point source emits the reference beam which is positioned in the front focal plane of the lens. Let $\delta(x+a,y)$ be the complex wave field of the wave leaving the point source, where $\delta(x,y)$ is the Dirac delta function and a is the distance between the object and the source, then the complex wave field of the reference beam at the photographic plate is given by:

$$\Delta(\xi,\psi) = F\{\delta(x+a,y)\} = \exp(-j2\xi a). \qquad (6.16)$$

The interference pattern irradiance produced by these two waves is then given by:

$$I(\xi,\psi)=1+|u(\xi,\psi)|^2+u(\xi,\psi)\exp(j2\pi\xi\, a)+u^*(\xi,\psi)\exp(-j2\pi\xi\, a)\,. \quad (6.17)$$

In order to reconstruct the image the hologram is placed in the front focal plane of the lens, as it is presented in figure 6.8., and illuminated with help of a collimated beam of monochromatic light. Let the incident wave has unit amplitude and that the transmittance of the hologram is a linear function of the irradiance, then the complex wave field of the transmitted way is equal to:

$$T(\xi,\psi)=t_0+\alpha\, I(\xi,\psi)\,. \quad (6.18)$$

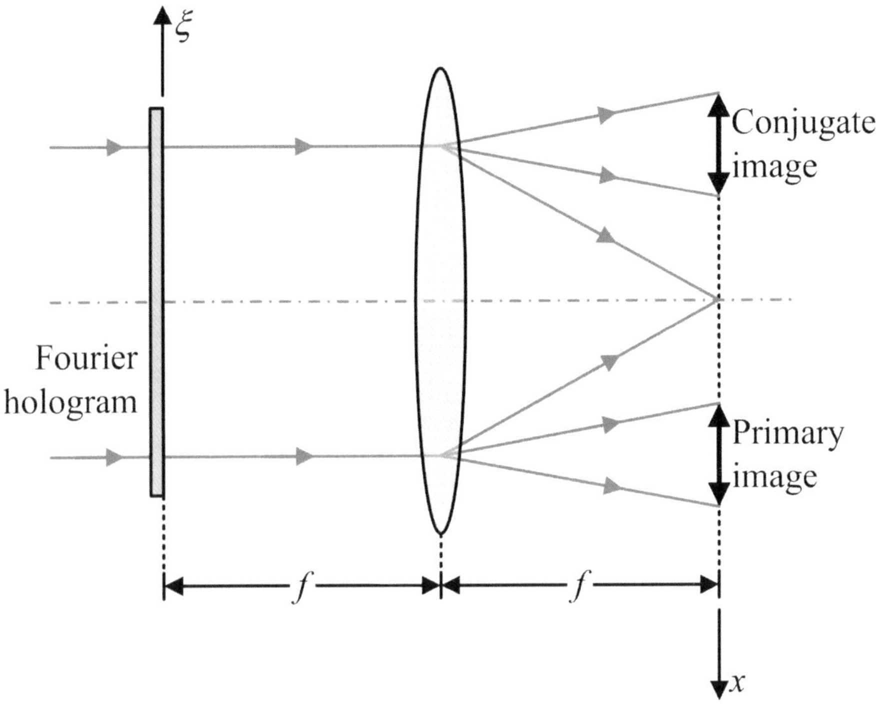

Figure 6.8. Fourier hologram image reconstruction.

The obtained complex wave field in the back focal length of the lens id the Fourier transform of the hologram transmitted wave, i.e.:

$$t(x,y)=F\{T(\xi,\psi)\}=(t_0+\alpha)\delta(x,y)+\alpha\, U(x,y)\otimes U(x,y)+ \\ +\alpha\, U(x-a,y)+\alpha\, U^*(-x+a,-y)\,, \quad (6.19)$$

where $\otimes$ represents the convolution operator. Figure 6.8. shows that the wave, which corresponds to the first term of the righthand side of equation (6.19), is focused on the optical axis where the second term represents a halo around it. The image of the original object is formed by the third term but shifted downwards by a distance a. Finally, the fourth term produces a conjugate image rotated by 180° and shifted upwards by the same distance a.

An interesting feature of the Fourier hologram is that the image does not move when the hologram changes it position in its own plane. The reason for this is that a function shift in the spatial domain results only that its Fourier transform is multiplied by a phase factor that consequently doesn't influence the irradiance distribution.

LENSLESS FOURIER HOLOGRAMS

Fourier hologram, with the same properties as one shown in the previous chapter, can be recorder but without using lens as it is presented in figure 6.9. In the presented setup the object is illuminated with a plane wave and as the source of the reference wave point source has been used which is positioned in the plane of the object.

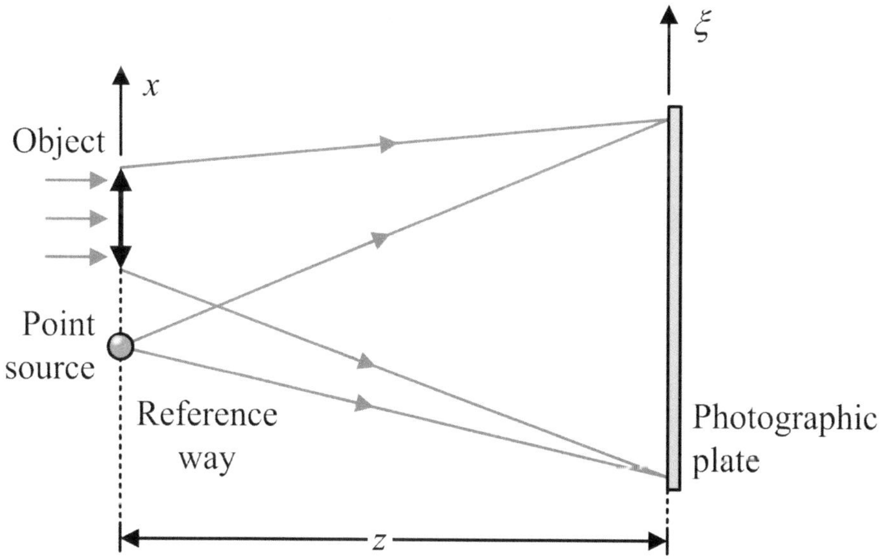

Figure 6.9. Fourier hologram recording optical system without using lens.

This optical arrangement eliminates the effect of the spherical phase factor associated with the Fresnel diffraction pattern by using the spherical reference wave with the same mean curvature.

REFERENCES

1. P. Hariharan. *Basics of Holography*. Cambridge: Cambridge University Press, 2002.
2. Ulf Schnars and Werner Jueptner. *Digital Holography: Digital Hologram Recording, Numerical Reconstruction, and Related Techniques*. Berlin: Springer-Verlag Berlin Heidelberg, 2005.
3. Kevin Curtis, Lisa Dhar, Adrian Hill, William Wilson, and Mark Ayres. *Holographic Data Storage: From Theory to Practical Systems*. Chichester: John Wiley & Sons Ltd, 2010.
4. Pierpaolo Boffi, Davide Piccinin, and Maria C. Ubaldi (Eds.). *Infrared Holography for Optical Communications: Techniques, Materials, and Devices*. Berlin: Springer-Verlag Berlin Heidelberg, 2003.
5. Ting-Chung Poon. *Optical Scanning Holography with MATLAB®*. New York: Springer Science+Business Media, LLC, 2007.
6. Chandra S. Vikram. *Particle Field Holography*. Cambridge: Cambridge University Press, 2005.
7. Chung J. Kuo and Meng Hua Tsai (Eds). *Three-Dimensional Holographic Imaging*. New York: John Wiley & Sons Inc., 2002.

7

CHAPTER

FIBER OPTICS

In the last two decades we are the witnesses of almost exponential growth of Internet traffic. The accessibility of generating and transferring of the multimedia content over Internet to an average customer on a daily basis pushes hardly the telecom companies to provide telecom networks which are capable of transmitting the data with very high capacity at relatively long distances. The technology that is capable to fulfill such stringent requirements is based on a network technology that comprises the use of an optical fiber as the signal propagating medium. Nowadays, the standard telecom optical fibers are capable of transferring the data with the capacity that exceeds 400 Gb/s per fiber which are based on the *DWDM (**D**ense **W**avelength **D**ivision **M**ultiplexing)* network technology. Baring in mind, that the standard single-mode fiber-optical cable, which is commonly manufactured, consists of maximum 864 fibers where the fibers are placed in 36 ribbons each containing 24 strands of fibers, the capacity per optical cable can reach a few hundreds of Tb/s. Moreover, the distances of 100 km can be reached between two optical nodes with an unamplified and uncompensated system with a bit rate of around 10 Gb/s. Therefore, the presented capabilities of the fiber-optic based telecommunication networks will further push the use of optical fibers to be the backbone of the future data transfer systems.

Although optical fibers are mostly used in the telecommunication networks for the high capacity data transfer, there is also another specific use of the optical fibers. Namely, having in mind that optical intefreometry is one of the most sensitive measuring techniques that is capable of high-precision and high-speed measurements of different physical quantities, a large number of different sensors have been developed which are based on the optical fibers. The fiber-optic based sensors, which are capable of measuring different physical as well as chemical parameters, such as temperature, strain, force, pressure, pH, concentration, index of refraction, etc. can be organized in such a way that they can be positioned within the fiber, i.e. to be the integral part of the optical fiber, or to be positioned on the tip of the fiber end. It is common, that based on the channeled spectrum analysis of the sensor back reflected light, we can arrange even a lab on a fiber tip where simultaneously can be interrogated several different physical and chemical quantities. Bearing in mind that optical fibers are passive device, which do not require electrical power supply, that are immune to the environmental electromagnetic radiation, and that they have zero electromagnetic radiation emission, they represent an ideal solution for different sensing applications in the hazard environments as well as in potentially explosive areas.

Sensors based on the optical fibers offer the possibility of simple multiplexing a large number of sensors even along a single optical fiber. In this way one can arrange quasi-distributed or even distributed measurements of different physical and chemical parameters. Therefore, the fiber-optical sensors are one of the most frequent choices when the structural integrity of different objects such as bridges, tunnels, large buildings, etc. must be permanently monitored.

All the above mentioned shows that the optical fiber is one of the most important optical component, which significantly changed our lives in the last couple of decades enabling high capacity data transfer to the every person whether it uses a standalone or a mobile device with the Internet access. This is the reason why the whole chapter is dedicated to the optical fibers.

OPTICAL FIBERS

At the very beginning of development of the optical waveguides, whose were aimed for the light transmission, it was very well known that the microwaves can be easily transmitted through the very simple waveguides. However, the problem with optical waveguides was the very high absorption of the optical materials which were used at the time thus severely limiting the transmission range. The researchers tested different but among the best suited was the glass known as fused silica consisting of silica in amorphous non-crystalline form. Even fused silica attenuated the light by at least one third after a distance no longer than 1 m that severely limited the transmission over the long distances.

The significant breakthrough in developing low loss optical fibers occurred in 1966 when the researchers became aware that the losses within the optical fibers are caused by the material impurities and not by the material itself. At that time, it was predicted that a glass based optical fibers can be produced with the attenuation instead of typical 1000 dB/km of "only" 20 dB/km. Since that time the optical losses of the optical fibers was constantly reducing reaching nowadays 0.2 dB/km.

The optical fibers can be roughly divided into two main groups, i.e. in the multimode and in the single-mode fibers. The multimode fibers are characterized by a relatively large diameter of the core through which the light is guided. The dimeter of the multimode core is usually much larger than the wavelength of the light as presented in figure 7.1. Typically, the core diameter is 50 μm embedded in a fiber cladding with the outside diameter of 125 μm. In contrast to the multimode fibers, the single-mode fibers have a core diameter that is slightly larger than the wavelength with the typical values ranging between 7 and 10 μm. However, the cladding diameter of the single-mode fibers is of the same value of 125 μm as in the case of the multimode fibers, which is also presented in figure 7.1.

The first applications of the optical fibers were reserved for the multimode type. This type of the optical fibers allow better mutual coupling efficiency of the fibers as there were fewer losses when connecting fibers together. In comparison with the multimode fibers the single-mode fibers allow higher data transmission rates over longer distances. Due to the small core diameter of the single-mode fibers the connector tolerance must be very tight in order to keep coupling losses low. At the beginning this was the limiting factor in using single-mode fibers. As the problem with the connector tolerance was solved satisfactorily, single-mode fibers became the favored choice, especially for the long hauls. Nowadays, the multimode fibers are used only for the short distances in particular in the local area networks.

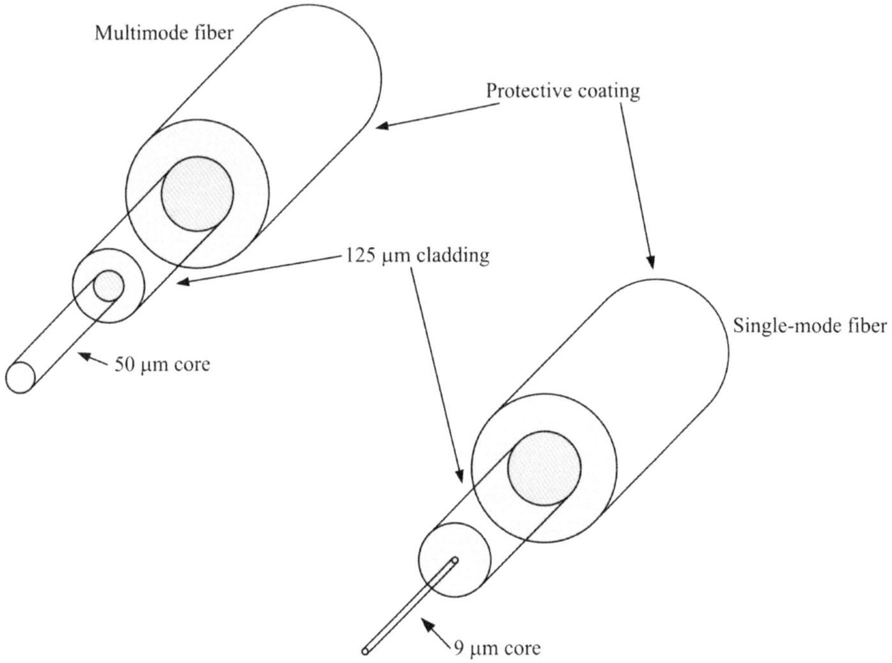

Figure 7.1. Multimode fibers and single-mode fibers dimensions.

Geometrical Optics and the Optical Fibers

In order to determine the light transmission parameters of an optical fiber there are two standard ways. The first one is based on the ray tracing technique, which is in the frame of the geometrical optics and which will be presented in this chapter. The second one considers the wave nature of the light where the Maxwell's equations in the cylindrical coordinate system will be used in order to treat the light propagation along an optical fiber.

The main physical principle behind the light propagation within an optical fiber in the frame of geometrical optics treatment is the well-known phenomenon of the total internal reflection that has been presented in Chapter 3. Typical optical fiber has a step index change of the index of refraction along its cross-section. There is a core with a circular cross-section which is surrounded by a cladding with a ring-shaped cross-section. The core is made of a glass with slightly higher refractive index than that of the cladding. Due to the total internal reflection phenomenon the light is guided through the core. If instead of a two-layer structure we have a single layer optical fiber, i.e. simple fiber without internal structure, the guidance is also possible. However, any dirt on the fiber surface or coating that should protect the fiber can produce that some rays that propagate inside the unstructured fiber leak out leading to the very high loses. In this case, i.e. when the dirt comes into contact to the unstructured fiber or any kind of a material with a similar index of refraction as of the fiber, the condition for the total internal reflection is broken down giving a rise to the leakage of the rays from the fiber and thus to the attenuation of the optical signal. By introducing the internal fiber structure the light guidance is not any more dependent on the fiber surroundings.

As the outer fiber surface is not any more important for guiding the light in step index fiber, we can, without losing the generality of the analysis, assume that the cladding diameter is infinitely wide, i.e. we can assume that there is no outside surface. In order to determine the largest possible angle, with respect to the fiber axis, that provides the condition for the ray propagation along the fiber we will observe the fiber cross-section as presented in figure 7.2.

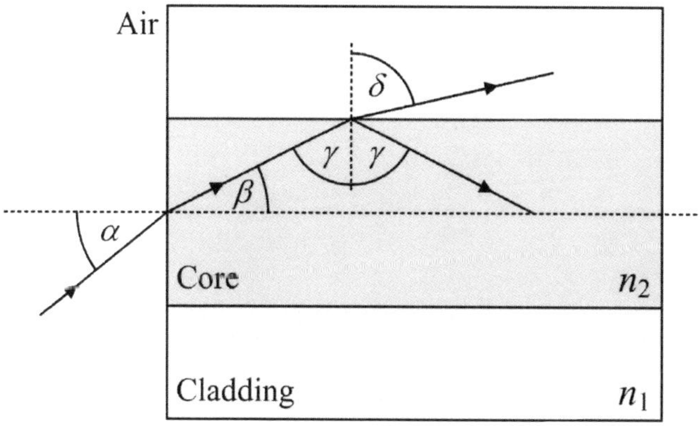

Figure 7.2. Cross-section of the step index fiber.

According to the Snell's law we have:

$$n_0 \sin\alpha = n_2 \sin\beta$$
$$n_2 \sin\gamma = n_1 \sin\delta \quad , \qquad (7.1)$$

where n_0 is the air index of refraction, n_1 is the cladding index of refraction, and n_2 is the core index of refraction, where the following is fulfilled $n_0 \le n_1 \le n_2$. If the fiber axis to be perpendicular to the front face we have $\beta+\gamma = \pi/2$, and consequently:

$$\sin\beta = \sqrt{1-\sin^2\gamma} \quad . \qquad (7.2)$$

The critical angle for the total internal refraction at the core-cladding interface is given by:

$$\delta_C = 90^{\circ} \quad \Rightarrow \quad \sin\delta_C = 1 \quad \Rightarrow \quad \sin\gamma_C = \frac{n_1}{n_2} \quad . \qquad (7.3)$$

The combination of the equations (7.1), (7.2) and (7.3) gives:

$$n_0 \sin\alpha_C = n_2\sqrt{1-\left(\frac{n_1}{n_2}\right)^2} = \sqrt{n_2^2 - n_1^2} \quad . \qquad (7.4)$$

If we further take $n_A = 1$ for air then the critical angle α_C for rays to be guided within the fiber core is:

$$\alpha_C = \arcsin\sqrt{n_2^2 - n_1^2} \quad . \qquad (7.5)$$

The argument of the inverse sine function has a special name, i.e. it is called the numerical aperture, which is often abbreviated as NA and given by:

$$NA = \sqrt{n_2^2 - n_1^2} \quad . \qquad (7.6)$$

It is obvious from equation (7.6) that the numerical aperture is a measure of the core and cladding indices of refraction difference. The largest angle of acceptance for rays impinging the fiber end facet is given by:

$$\alpha_C = \arcsin NA \quad . \qquad (7.7)$$

This angle represents at the same time the angle of the light exit cone at the fiber end when the light exits an optical fiber. The corresponding input/output cone is schematically shown in figure 7.3.

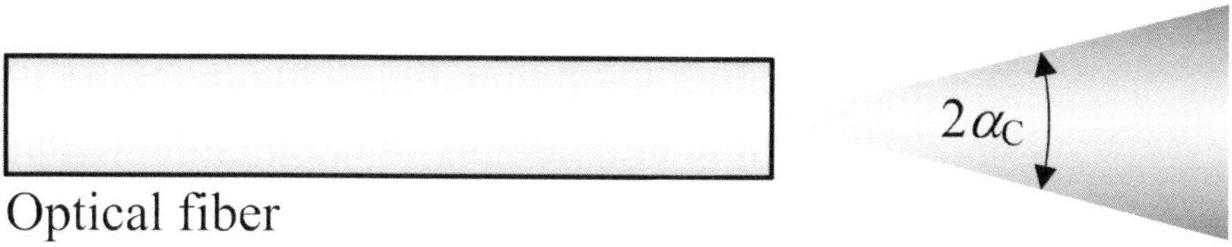

Figure 7.3. Acceptance and exit cone of an optical fiber.

There is another parameter which is also frequently used when speaking about an optical fiber. This parameter is given by:

$$\Delta = \frac{n_2^2 - n_1^2}{2n_2^2} \quad . \qquad (7.8)$$

The relation between NA and the Δ parameter is:

$$NA = n_2\sqrt{2\Delta} \quad . \qquad (7.9)$$

Typically, the values of core and cladding indices of refraction are very close to each other with a mutual difference of a few tenths of 1%. Bearing this in mind equation (7.8) becomes:

$$\Delta \approx \frac{(n_2 + n_1)(n_2 - n_1)}{n_2(n_2 + n_1)} \approx \frac{n_2 - n_1}{n_2} \quad . \qquad (7.10)$$

According to the last relation it is obvious why the parameter Δ is called normalized refractive index difference or normalized index step. For a typical single-mode fiber we have Δ = 0.35% and n_2 = 1.45 which leads to NA = 0.12, which according to equation (7.7) gives the value of ±6.9° for the acceptance angle.

As it was mentioned above, all the rays that enter the fiber end facet within the acceptance cone will propagate throughout the fiber core. The rays travel along the fiber bouncing between the core-cladding interfaces with different angles thus traveling the different path lengths until they reach the far end of the fiber. Consequently they arrive at different times and give a rise to the modal dispersion which is illustrated in figure 7.4.

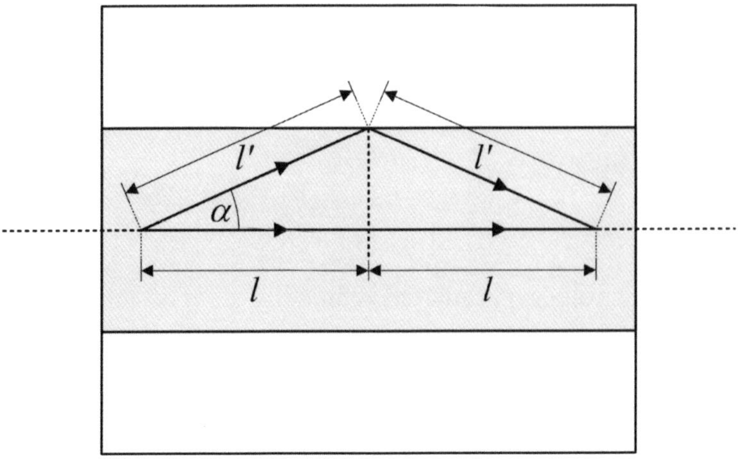

Figure 7.4. Modal dispersion.

According to the ray paths, shown in figure 7.4., one can noticed that the ray that impinge the core-cladding interface and travels with the angle α between the ray and the fiber axis, travels the path which is larger than the ray which travels along the fiber axis and where it is clearly valid $l' = l/\cos\alpha$. It was previously shown that $\sin\alpha \leq NA/n_2$ is valid, so if it is fulfilled α << 1, we have $\sin\alpha \approx \alpha$ and $\cos\alpha \approx 1 - \alpha^2/2$, and consequently:

$$l' \approx l\left(1 + \frac{NA^2}{2n_2^2}\right) \approx l(1+\Delta) \quad . \qquad (7.11)$$

As the optical signal that propagate through the fiber travels along different paths there is a scatter of propagation times which is often called delay distortion. When the optical signal is in the shape of the short pulses, the pulse propagation causes a widening of the pulse. This is not the problem until the pulse becomes wider than the pulse rate. When this happen, an intersymbol interference occurs and the transmitted message becomes undecipherable. In order to show that this can be a serious problem a rough calculation will be performed. If we assume that the light velocity within the fiber core is c/n_2 then the propagation time of the light in the fiber with the length L is $\tau = n_2L/c$. According to the modal dispersion geometry shown in figure 7.4., for the minimal time travel we have:

$$\tau_{min} = \frac{n_2 l}{c}, \qquad (7.12)$$

where the maximal time travel is reached for the maximal ray angle and is given by:

$$\tau_{max} = \frac{n_2 l'}{c} = \frac{n_2 l}{c}(1+\Delta). \qquad (7.13)$$

The difference in time travel is equal to:

$$ä\tau = \tau_{max} - \tau_{min} = \frac{n_2 l}{c}\Delta = \tau_{min}\Delta. \qquad (7.14)$$

The corresponding propagation time scatter is then given by:

$$\frac{ä\tau}{\tau_{min}} = \tau_{min}\Delta. \qquad (7.15)$$

If we take, as in the previous case, that $\Delta = 0.35\%$ then for a fiber length of 1 km we have a pulse broadening equal to $\delta\tau \approx 17.5$ ns. This represents only a rough estimation as we assumed that only meridional rays (rays that pass through the optical fiber axis) propagate along the fiber neglecting the existence of the helical rays (rays that bounce around the optical fiber axis) which are presented in figure 7.5. However, it is obvious that the modal dispersion can significantly broad the pulse which further limits the data rate that in our case is equal to the inverse value of the pulse broadening, i.e. $1/\delta\tau \approx 57$ Mb/s. One can notice that is very low data rate if compared with today's standard telecom data rates especially in our case where the fiber length of only 1 km was considered. The mechanism of modal dispersion can severely hamper the use of an optical fiber for practical applications. However, there are ways in order to avoid the problem with modal dispersion. One of them is to use the so-called gradient index fibers, but in the case of the highest data rates the single-mode fibers must be used.

In order to suppress the modal dispersion and thus to avoid the scatter of arrival times an optical fiber with a gradient radial profile of the refractive index must be used. Such optical fibers have the following radial profile of an index of refraction along the fiber cross-section:

$$n(r)=\begin{cases} n_2\sqrt{1-2\Delta\left(\dfrac{r}{a}\right)^{\alpha}}, & |r|\le a \\ n_1, & |r|>a \end{cases} \quad (7.16)$$

where a denotes the core radius and α is the corresponding exponential parameter. The resulting index of refraction profile for the given exponential parameter values is presented in figure 7.6.

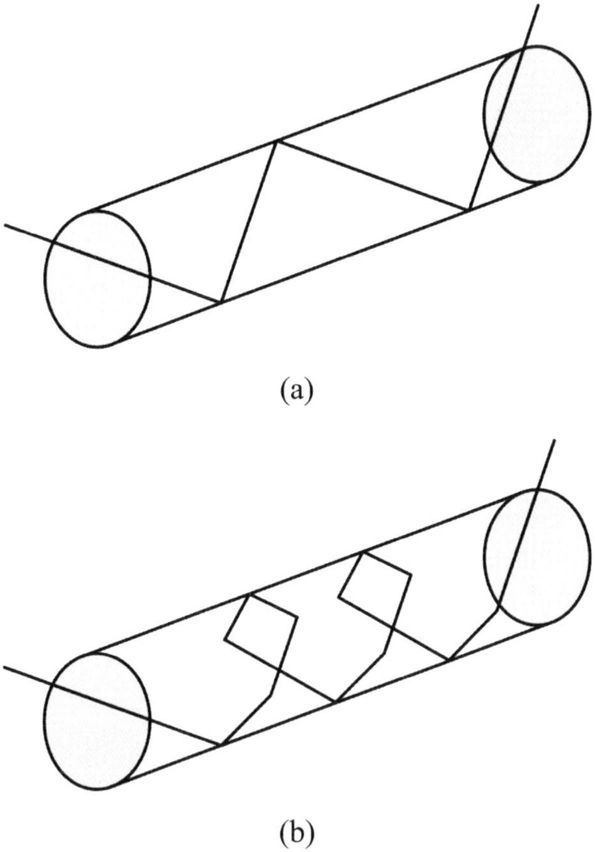

Figure 7.5. (a) Meridional and (b) helical rays in an optical fiber.

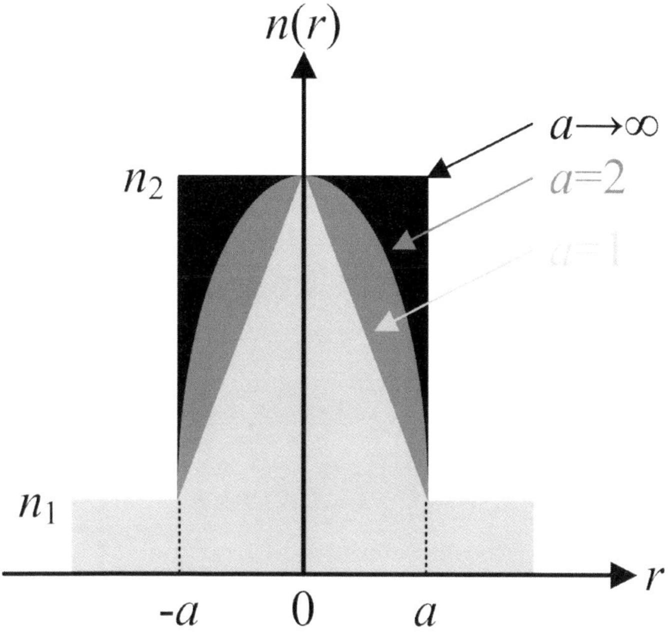

Figure 7.6. Some common index profiles of a gradient optical fiber.

In order to minimize the differences in transit time there is an optimal profile of an index of refraction. The first approximation gives us an optimal profile for α = 2, i.e. a parabolic index profile. The rays travel curved paths in the fiber and not zigzag paths as in the case of the step index fiber. Although the curved paths are longer than the straight paths along the fiber axis, i.e. the larger the detour of the ray from the axis the longer the ray paths, the index of refraction is smaller and thus the higher is the detour ray velocity. In this way longer paths are compensated by the higher ray velocities and consequently the corresponding optical paths are the same. The scatter of the transit time depends on the exponential factor and for a given values of the exponential factor we have the following:

$$\ddot{a}\,\tau = \begin{cases} \dfrac{n_2 l}{c}\Delta, & \alpha \to \infty \\ \dfrac{n_2 l}{c}\dfrac{\Delta^2}{2}, & \alpha = 2 \end{cases} \quad . \qquad (7.17)$$

If we compare the scatter times for α → ∞ and α = 2 we obtained in the case of the parabolic profile gradient fiber the reduction of the scatter time by the factor $\Delta/2 \approx 10^{-3}$, which represents a considerable improvement, where for the parabolic profiled gradient fiber the transit time spread is reduced by about three orders of magnitude. The modal dispersion is then consequently reduced to a few tens of picosecond per kilometers. Precise calculations show that the optimal profile is not obtained for $\alpha = 2$ but for a slightly different value. This optimal value typically depends on glass type, doping material, and wavelength. Furthermore, the optimum exponent value is not the same for meridional rays and helical rays. Therefore, it is difficult to theoretically predict the optimum value and thus its significance is reduced, especially if we take into consideration the unavoidable manufacturing tolerances during the process of the fiber productions.

Wave Optics and the Optical Fibers

The standard optical fiber structure implies rotational symmetry. Therefore, the Maxwell's equations, which have been introduces in Chapter 1. in the Cartesian coordinate system, must be adapted in order fits better the cylindrical structure of an optical fiber. This strongly suggests the use of a cylindrical coordinate system with the corresponding cylindrical coordinates r, φ, and z. It is very well known that the Laplacian in the cylindrical coordinates is given by:

$$\nabla^2 E = \frac{1}{r}\frac{\partial}{\partial r}\left(r\frac{\partial E}{\partial r}\right) + \frac{1}{r^2}\frac{\partial^2 E}{\partial \varphi^2} + \frac{\partial^2 E}{\partial z^2} \quad . \tag{7.18}$$

If we further introduce a separation of the variables in the following way:

$$E(r,\varphi,z,t) = E_0 \rho \zeta \tau \quad , \tag{7.19}$$

where E_0 is the wave field amplitude and where:

$$\rho = \rho(r,\varphi) \quad , \tag{7.20}$$

is the field amplitude distribution in the plane normal to the z-axis, i.e. across the fiber cross-section,

$$\zeta = \zeta(z) = \exp(-j\beta z) \quad , \tag{7.21}$$

where β is the corresponding fiber wave number, and:

$$\tau = \tau(t) = \exp(j\omega t) \quad , \tag{7.22}$$

represents a monochromatic wave with an angular frequency ω. The variable separation can be performed due to the linearity and paraxiality of the in fiber propagating rays, which further leads that both the electric and magnetic field components are in the first approximation perpendicular to the propagation direction. Therefore, the longitudinal and transverse processes are practically decoupled. Taking the above presented into consideration and equation (7.18) the corresponding wave equation becomes:

$$\frac{n^2}{c^2}\frac{\partial^2}{\partial t^2}(E_0\rho\zeta\tau)=\frac{1}{r}\frac{\partial}{\partial r}\left[r\frac{\partial}{\partial r}(E_0\rho\zeta\tau)\right]+\frac{1}{r^2}\frac{\partial^2}{\partial\varphi^2}(E_0\rho\zeta\tau)+\frac{\partial^2}{\partial z^2}(E_0\rho\zeta\tau) \qquad .(7.23)$$

The first term on the right-hand side of equation (7.23) is:

$$\frac{1}{r}\frac{\partial}{\partial r}\left[r\frac{\partial}{\partial r}(\rho\zeta\tau)\right]=\xi\ \frac{1}{r}\frac{\partial}{\partial r}\left(r\frac{\partial\rho}{\partial r}\right), \qquad (7.24)$$

where the constant factor of the wave amplitude E_0 was omitted as it cancels out in the equation (7.23) because all terms contain this factor. The second term on the right-hand side

$$\frac{1}{r^2}\frac{\partial^2}{\partial\varphi^2}(\rho\zeta\tau)=\xi\ \frac{1}{r^2}\frac{\partial^2\rho}{\partial\varphi^2}. \qquad (7.25)$$

The third term becomes:

$$\frac{\partial^2}{\partial z^2}(\rho\zeta\tau)=\rho\ \frac{\partial^2\zeta}{\partial z^2}=-\beta^2\rho\zeta\tau. \qquad (7.26)$$

The only term on the left-hand side of equation (7.23) is:

$$\frac{n^2}{c^2}\frac{\partial^2}{\partial t^2}(\rho\zeta\tau)=\beta\ \frac{n^2}{c^2}\frac{\partial^2\tau}{\partial t^2}=-\frac{n^2}{c^2}\omega^2\rho\zeta\tau. \qquad (7.27)$$

If the vacuum wave number is denoted as k_0 then in the medium with the index of refraction n we have $k = nk_0 = n\omega/c$, so equation (7.27) becomes:

$$\frac{n^2}{c^2}\frac{\partial^2}{\partial t^2}(\rho\zeta\tau)=-k^2\rho\zeta\tau. \qquad (7.28)$$

By inspecting equations (7.24), (7.25), (7.26), and (7.28) on can noticed that the factor ξ common to all terms and therefore it is cancelled out. Due to the rotational symmetry of the fiber cross-section we can further separate the variables in the following way:

$$\rho(r,\varphi)=R(r)\Phi(\varphi). \quad (7.29)$$

Equation (7.23) in this case becomes:

$$-k^2r^2=\frac{r}{R}\frac{\partial R}{\partial r}+\frac{r^2}{R}\frac{\partial^2 R}{\partial r^2}+\frac{1}{\Phi}\frac{\partial^2\Phi}{\partial\varphi^2}-\beta^2r^2. \quad (7.30)$$

The reorganization of the last equation gives:

$$-\frac{1}{\Phi}\frac{\partial^2\Phi}{\partial\varphi^2}=\frac{1}{R}\left[r\frac{\partial R}{\partial r}+r^2\frac{\partial^2 R}{\partial r^2}+r^2\left(k^2-\beta^2\right)R\right]. \quad (7.31)$$

The last equation contains on the left-hand side only the function $\Phi(\varphi)$, i.e. depends only on the angle φ, and where the right-hand site contains only the $R(r)$ function, i.e. depends only on the radius r. As both sides are equal, they must be constant, i.e. they don't depend either on φ and/or r. This constant will be denoted as p^2. In this case equation (7.31) can be divided into two equations in the following form:

$$\frac{\partial^2\Phi}{\partial\varphi^2}+p^2\Phi=0, \quad (7.32)$$

and:

$$r^2\frac{\partial^2 R}{\partial r^2}+r\frac{\partial R}{\partial r}+r^2\left(k^2-\beta^2\right)R=p^2R. \quad (7.33)$$

If we take into consideration the following substitution $\kappa^2=k^2-\beta^2$, equation (7.33) becomes:

$$r^2\frac{\partial^2 R}{\partial r^2}+r\frac{\partial R}{\partial r}+\left(\kappa^2r^2-p^2\right)R=0. \quad (7.34)$$

Partial differential equation, given by equation (7.32) has the general solution in the following form:

$$\Phi(\varphi)=\Phi_0\cos(p\varphi+\varphi_0), \quad (7.35)$$

where Φ_0 and φ_0 are the constants. Further, partial differential equation, given by equation (7.34) is in the form:

$$r^2R''+rR'+\left(\kappa^2r^2-p^2\right)R=0. \quad (7.36)$$

The solution of this differential equation is given in the form:

$$R(r) = \rho_1 J_p(\kappa r) + \rho_2 N_p(\kappa r), \tag{7.37}$$

where κr is real ($\kappa^2 r^2 \geq 0$), or:

$$R(r) = \rho_3 I_p(\kappa r) + \rho_4 K_p(\kappa r), \tag{7.38}$$

where κr is imaginary ($\kappa^2 r^2 < 0$). The functions $J_p(x)$, $N_p(x)$, $I_p(x)$, and $K_p(x)$ are the corresponding Bessel functions.

In order to determine the constants ρ_1, ρ_2, ρ_3, and ρ_4 we must specify the geometry of the used optical fiber. Since this moment we have only assumed that fiber is of a cylindrical shape and that the core and cladding indices of refraction are just slightly different. Further on, we will consider a step index fiber with the following index of refraction distribution:

$$n = \begin{cases} n_2, & r \leq a \\ n_1, & r > a \end{cases}, \tag{7.39}$$

where it is $n_2 > n_1$ taken into account as otherwise there would be no wave propagation with small loses. In the case when the core radius is much larger than the wavelength of the propagating light the obtained solution must follows the results obtained in the frame of the geometrical optics where the ray tracing technique was taken into account, i.e. the light is guided within the fiber core and the solution must predict that the most of the light wave is concerted in the fiber core. Also, the solution must provide that the light within the cladding will have the amplitude that decreases with the radius increase. Therefore, the parameter κr must be real in the core and imaginary in the cladding, i.e. $k \geq \beta$ must be fulfilled for the core, where the opposite is valid for the cladding. Further, in order to have parameter κr real in the core one must fulfill $\kappa^2 r^2 \geq 0$ or $(k_2^2 - \beta^2)r^2 \geq 0$ for $r \leq a$, where k_2 is the core wave number. The case is opposite for the cladding where the following must be fulfilled $\kappa^2 r^2 < 0$ and consequently $(k_1^2 - \beta^2)r^2 < 0$ for $r > a$, where k_1 is the cladding wave number. Following the last two inequalities we have:

$$k_1 < \beta < k_2. \tag{7.40}$$

For the transversal components of the core and cladding wave number the following is valid:

$$\kappa_2^2 = k_2^2 - \beta^2$$
$$\kappa_1^2 = -\left(k_1^2 - \beta^2\right), \qquad (7.41)$$

respectively. It is further useful to make the following substitutions:

$$u = a\kappa_2$$
$$w = a\kappa_1, \qquad (7.42)$$

where u and w are dimensionless, real positive quantities which are often used in the literature. The parameter u defines the progression of the phase and the parameter w defines the transverse decay of the wave amplitude. The following relations are valid for these two parameters:

$$\begin{aligned} u^2 + w^2 &= a^2\left(k_2^2 - \beta^2\right) - a^2\left(k_1^2 - \beta^2\right) \\ &= a^2\left(k_2^2 - k_1^2\right) \\ &= a^2 k_0^2\left(n_2^2 - n_1^2\right) \end{aligned}, \qquad (7.43)$$

so the parameter $u^2 + w^2$ represents a constant. This constant is of very high importance in the fiber optics and it is called normalized frequency or simply V number, which is given by:

$$V^2 = u^2 + w^2 = a^2 k_0^2\left(n_2^2 - n_1^2\right), \qquad (7.44)$$

or equivalently:

$$\begin{aligned} V &= k_0 a NA \\ &= \frac{2\pi}{\lambda_0} a\sqrt{n_2^2 - n_1^2} \end{aligned}. \qquad (7.45)$$

By using the parameters u and w one can obtain the general solution for the step index fiber in the following form:

$$\rho_2(r,\varphi) = C_2 J_p\left(\frac{u}{a}\right)\cos(p\varphi + \varphi_0)$$
$$\rho_1(r,\varphi) = C_1 K_p\left(\frac{u}{a}\right)\cos(p\varphi + \varphi_0), \qquad (7.46)$$

where the first solution in equation (7.45) is valid for the core (subscript 2) and the second one for the cladding (subscript 1). In order to fulfill the boundary conditions for $r = a$ both solutions given in equation (7.46) must have the same value, so the following must be fulfilled:

$$\rho_2(r=a,\varphi)=\rho_1(r=a,\varphi)$$
$$\frac{\partial}{\partial r}\rho_2(r=a,\varphi)=\frac{\partial}{\partial r}\rho_1(r=a,\varphi) \quad (7.47)$$

where after substituting equations (7.46) into (7.47) we have:

$$C_2 J_p(u)=C_1 K_p(w)$$
$$C_2\frac{\partial}{\partial r}J_p\left(\frac{u}{a}\right)\Bigg|_{r=a}=C_1\frac{\partial}{\partial r}K_p\left(\frac{w}{a}\right)\Bigg|_{r=a} . \quad (7.48)$$

Based on the following identity:

$$\frac{\partial}{\partial r}=\frac{u}{a}\frac{\partial}{\partial(u/a)}=\frac{w}{a}\frac{\partial}{\partial(w/a)} , \quad (7.49)$$

the second relation in equation (7.49) becomes:

$$C_2\frac{u}{a}J'_p(u)=C_1\frac{w}{a}K'_p(w) , \quad (7.50)$$

where the prime denotes the first derivative with respect to the argument. In order to exists the solutions for the constants C_1 and C_2, the following relation must be fulfilled (the determinant of the linear system of equation must be zero):

$$C_1C_2\frac{u}{a}K_p(w)J'_p(u)-C_1C_2\frac{w}{a}J_p(u)K'_p(w)=0 . \quad (7.51)$$

It is well known that for the Bessel functions is valid:

$$uJ'_p(u)=pJ_p(u)-uJ_{p+1}(u)$$
$$wK'_p(w)=pK_p(w)-wK_{p+1}(w) . \quad (7.52)$$

The combination of equations (7.51) and (7.52) finally gives:

$$\frac{J_p(u)}{uJ_{p+1}(u)}=\frac{K_p(w)}{wK_{p+1}(w)} . \quad (7.53)$$

From the last equation one can obtain the solutions for the optical fiber fundamental modes. From the Bessel functions properties it is obvious that the right-hand side of equation (7.53) is always positive while the left-had side frequently change its sign. Therefore, there are certain values of the arguments u and w, and consequently certain values of the V number where the solution is possible.

In the following we will analyze the case when $p = 0$ which is valid for the rotationally symmetric field across the fiber's cross-section. Equation (7.53) in this case becomes:

$$\frac{J_0(u)}{uJ_1(u)} = \frac{K_0(w)}{wK_1(w)}. \tag{7.54}$$

By analyzing the equation (7.54) one can notice that there are range alternations where the solutions can and cannot be found. The corresponding graphs of the left-hand and the right-hand side of the equation (7.54) are presented in figure 7.7. In order to be able to find the solutions, both sides of the equation (7.54) must have the same sign. Therefore, and according from the figure 7.7., the solutions exist if $0 \leq u \leq 2.405$, $3.832 \leq u \leq 5.520$, etc., i.e. the ranges where the solutions are possible are marked with the gray color.

In the case when it is satisfied $p = 1$, according to equation (7.53) we have:

$$\frac{J_1(u)}{uJ_2(u)} = \frac{K_1(w)}{wK_2(w)}. \tag{7.55}$$

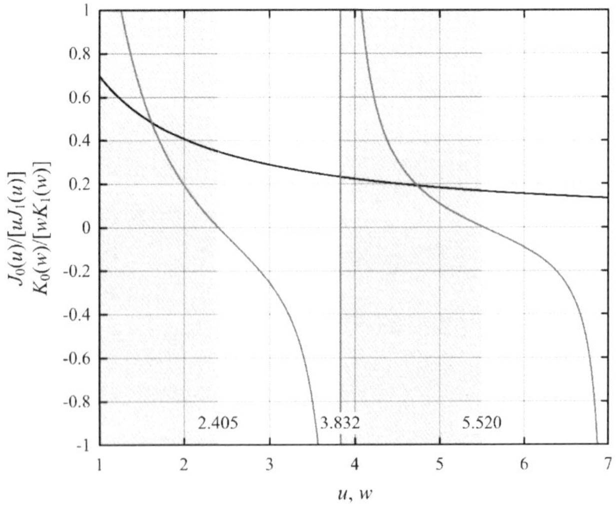

Figure 7.7. Graphs of the left-hand and the right-hand side of the equation (7.54).

However, in order to refer to the case when $p = 0$ is satisfied, we will start from the equation (7.52), but write it down in a slightly different recurrence form as:

$$uJ'_p(u) = -pJ_p(u) - uJ_{p-1}(u)$$
$$wK'_p(w) = -pK_p(w) - wK_{p-1}(w) \quad (7.56)$$

If we combine the last equation with equation (7.51) we obtain:

$$-\frac{J_1(u)}{uJ_0(u)} = \frac{K_1(w)}{wK_0(w)} \quad (7.57)$$

By analyzing the equation (7.57) one can also notice, like in the previous case, that there are range alternations where the solutions can and cannot be found. The corresponding graphs of the left-hand and the right-hand side of the equation (7.57) are presented in figure 7.8. In order to be able to find the solutions, both sides of the equation (7.57) must have the same sign. Therefore, and according from the figure 7.8., the solutions exist if $2.405 \leq u \leq 3.832$, $5.520 \leq u \leq 7.016$, etc., i.e. the ranges where the solutions are possible are marked with the gray color.

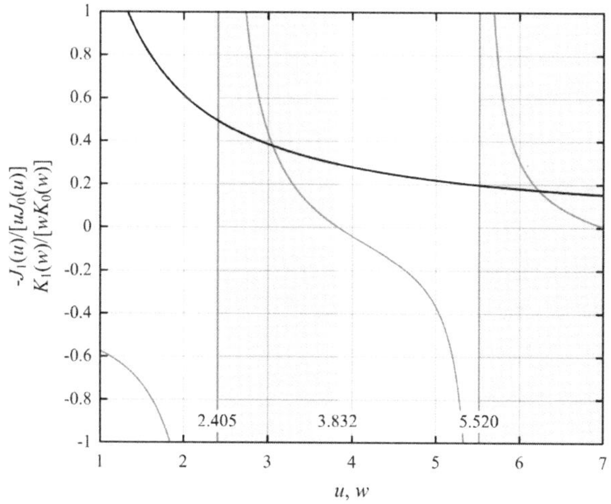

Figure 7.8. Graphs of the left-hand and the right-hand side of the equation (7.57).

Similar situation is for larger values of the parameter p, i.e. the ranges with and without solutions are alternating. What can be concluded from the previous analysis is that for $V < 2.405$, there is only one branch of solution and for $V \geq 2.405$ there are initially two branches. The higher the value of the V number the more branches come up.

The value of the V number of 2.405 represents the breaking point where there is more than one solution of the wave field distribution across the fiber cross-section. Below this value there is only one wave field distribution which can propagate along the fiber so it is said for the fiber to be single-moded. This first mode is called the fundamental mode. Depending on the fiber geometry and index of refraction distribution along its cross-section one can calculate the fiber cutoff wavelength from the condition $V = 2.405$. If the wavelength of the propagating light is longer than the cutoff value the fiber is single-moded.

In the case of so called multimode fibers, i.e. the fibers where beside the propagation of the fundamental mode there are also present higher modes. Each mode is formed from a set of two parameters. One of the parameters is p which indicates the angular dependence of the wave field distribution of the mode. If we have $p = 0$ the distribution is rotational invariant, i.e. it has rotational symmetry. In the case where $p = 1$ is fulfilled the wave field amplitude has a sine function dependence of the azimuth angle. Therefore, there are two zeroes at opposite positions, i.e. zeroes are located along the diameter. Between the zeros there are a positive and a negative branch or lobe. In the branches there are located the wave field irradiance maximums. As the sign of the field phases are opposite in the lobes, the wave fields oscillate with the opposite phase. Similarly for $p = 2$ there is four lobes as we have two full periods of the sine function around the fiber cross-section. In general we have $2p$ lobes for any value of the parameter p.

Parameter p designates which Bessel function describes the wave field distribution along the radial direction. It was found that Bessel function of the first kind J_p governs the field distribution within the core and the modified Bessel function of the second kind K_p governs the field distribution within the cladding. Bessel function of the first kind oscillates so there is unlimited number of ways to smoothly connect J_p to K_p for a given p. This set of possible solutions is marked with the parameter q, i.e. the second parameter. Each possible mode is designated with "LP_{pq}" as they are essentially linearly polarized where the parameter p represents the number of node pairs in the azimuth angle, and the parameter q represents the number the possibilities along the radial coordinate.

In order to present the possible irradiance distribution within the fiber, figures 7.9., 7.10., and 7.11. show the various possibilities to connection the functions J_p and K_p in a right way and give an idea about the irradiance distribution.

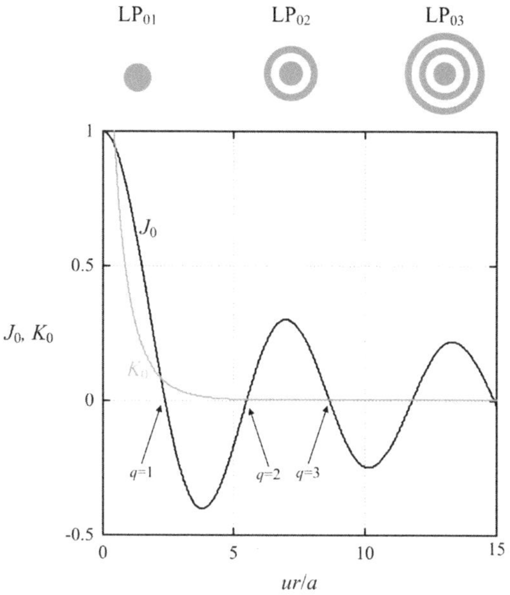

Figure 7.9. Radial irradiance distribution for modes with $p = 0$.

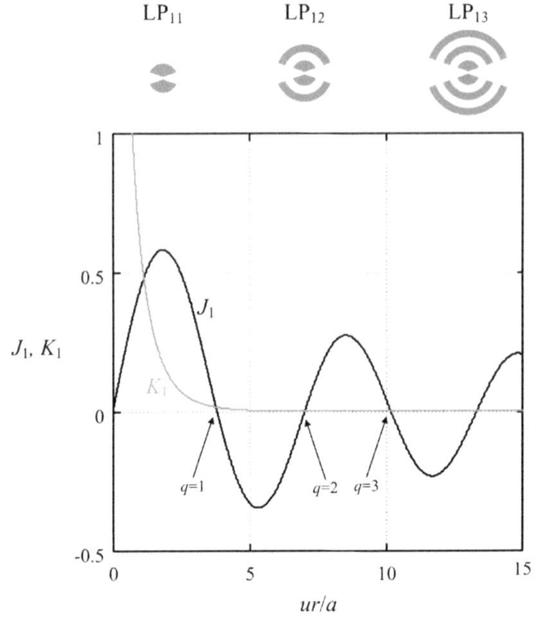

Figure 7.10. Radial irradiance distribution for modes with $p = 1$.

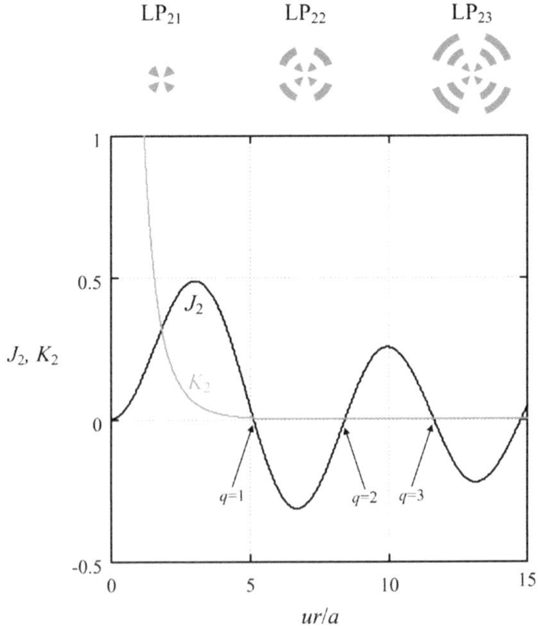

Figure 7.11. Radial irradiance distribution for modes with $p = 2$.

In order to illustrate the transition from a single-mode to a multimode fiber we will consider a typical optical fiber with the following set of parameters: the fiber core radius $a = 4$ μm, the core index of refraction $n_2 = 1.45$, and the index difference between core and cladding $\Delta = 0.35\%$. It is well known that the fiber numerical aperture is given by $NA = n_2\sqrt{2\Delta}$, where further from the cutoff condition $V = 2.405$ and the definition of the V number, for the cutoff wavelength condition we have:

$$\lambda_C = \frac{2\pi a n_2\sqrt{2\Delta}}{2.405}. \qquad (7.58)$$

By inserting the corresponding parameter values we obtain $\lambda_C = 1.268$ μm. Therefore, this fiber is a single-mode fiber for all wavelengths longer than 1.268 μm, which then includes the 1.3 μm and 1.55 μm range which is preferred in telecommunications. For the shorter wavelengths the fiber is then multimode. The second mode starts to propagate for the cutoff wavelength which is obtained according to equation (7.58) where the value of $V = 2.405$ is changed with the value $V = 3.832$. One obtains the cutoff wavelength of 796 nm for the second mode. For the larger V number values, i.e. the shorter wavelengths, one obtains more and more modes which can propagate.

For a given value of the V number there is a certain number of modes that can propagate along the fiber. In the case of large V number there is an approximate formula which estimates the number of mode as:

$$M \approx \frac{V^2}{2}, \qquad (7.59)$$

for a step index fiber. For a gradient index fiber it is slightly different:

$$M \approx \frac{V^2}{4}. \qquad (7.60)$$

Chromatic Dispersion

The dependence of the fiber core and cladding indices of refraction on the wavelength can affect the light propagation along the fiber. In the previous chapter we have dealt with the phenomenon of modal dispersion where it was noticed that different modes travel with different velocities thus giving a rise to a modal dispersion phenomenon. The modal dispersion can severely affect the data transport at higher rates thus limiting the maximal allowable data rates where the signal deciphering is still achievable. Similar situation is if a multimode optical fiber is deployed in some kind of a measurement instrument which is based on a phase sensitive detection, i.e. the interferometer base measurement system. Typically, if the higher modes are present in the fiber-optic interferometer the coherence length of the fiber propagating light is thus reduced which further can affect the measurement if the interferometer branches lengths are misbalanced.

Single-mode fibers successfully overcome the problem of the modal dispersion. However, there is still a chromatic dispersion which is also present in the single-mode fibers which can significantly influence the characteristic of the fiber-optic data links at very high rates. Therefore, the phenomenon of the chromatic dispersion will be analyzed in the following. The cause of the chromatic dispersion is, certainly the very well-known phenomenon of the index of refraction of the glass material from which the fiber is made of dependence on the wavelength. As there is no ideally monochromatic light, i.e. each signal's and thus the optical signal's spectrum is consisted of at least a very narrow range of frequencies, the chromatic dispersion occurs. Due to the limited spectral width different spectral components propagate with different velocity giving a rise to differential transit time at the fiber end and thus to signal distortion. There are three contributing factors which are hidden behind the wavelength dependence of the refractive index collectively called chromatic dispersion:

- Material dispersion, which contribution comes directly from the wavelength dependence of the glass material index of refraction. This type of dispersion is not specific to fibers only but can be generally found in any bulk glass and it is independent on the geometry.
- Waveguide dispersion, which contribution comes from the particular geometry of the optical fibers where the ratio of optical powers that propagate through the fiber core and cladding depends on the wavelength. As the light spectrum has different wavelengths, we have a crossover from mostly core index to mostly cladding index, which results in a contribution to the wavelength dependence of the effective index.

- Profile dispersion, which contribution comes from the normalized refractive index difference dependence on the wavelength. The main cause of this type of dispersion is that the core and cladding index of refraction do not vary in the same way. However, this type of dispersion is often much smaller than the material and waveguide dispersion.

In order to characterize the chromatic dispersion it is useful to specify the amount of the effect per unit distance. Therefore, in general case, for the chromatic dispersion, as well as for the modal and polarization mode dispersion we can write:

$$D = \frac{1}{L} \text{ä} \tau \quad . \qquad (7.61)$$

where δτ represents the difference of propagation time after fiber length L. Units that is commonly used is ps/km. In the case of the chromatic dispersion, which includes material, waveguide, and profile dispersion, we have:

$$D = \frac{1}{L} \frac{\mathrm{d}\tau}{\mathrm{d}\lambda} , \qquad (7.62)$$

where the parameter D contains of three components:

$$D = D_{\mathrm{M}} + D_{\mathrm{W}} + D_{\mathrm{P}} , \qquad (7.63)$$

representing the corresponding material D_{M}, waveguide D_{W} and profile D_{P} dispersions. The chromatic dispersion represents the propagation time difference per distance and per wavelength difference. Therefore, the unit that is commonly used is ps/(nm km).

If we consider a plane monochromatic wave with the angular frequency ω and the wave number β it propagates along the fiber with phase velocity given by:

$$v_{\mathrm{p}} = \frac{\omega}{\beta} , \qquad (7.64)$$

whereas the signal propagation velocity is governed by the group velocity given by:

$$v_{\mathrm{g}} = \frac{\mathrm{d}\omega}{\mathrm{d}\beta} . \qquad (7.65)$$

Therefore, the signal time of travel will be:

$$\tau = \frac{L}{v_{\mathrm{g}}} = L\frac{\mathrm{d}\beta}{\mathrm{d}\omega}$$
$$= L\frac{\mathrm{d}\beta}{\mathrm{d}\lambda}\frac{\mathrm{d}\lambda}{\mathrm{d}\omega} \quad . \tag{7.66}$$

Since we have:

$$\beta = \boldsymbol{k}_0 = \frac{2\pi n}{\lambda} , \tag{7.67}$$

it is valid:

$$\frac{\mathrm{d}\beta}{\mathrm{d}\lambda} = 2\pi\frac{\mathrm{d}}{\mathrm{d}\lambda}\left(\frac{n}{\lambda}\right) = \frac{2\pi}{\lambda^2}\left(\lambda\frac{\mathrm{d}n}{\mathrm{d}\lambda} - n\right) , \tag{7.68}$$

and also:

$$\frac{\mathrm{d}\lambda}{\mathrm{d}\omega} = -2\pi\frac{c}{\omega^2} \quad . \tag{7.69}$$

The combination of equations (7.66), (7.68), and (7.69) gives:

$$\tau = \frac{L}{c}\left(n - \lambda\frac{\mathrm{d}n}{\mathrm{d}\lambda}\right) . \tag{7.70}$$

Equation (7.70) represents the wavelength dependence of the group (signal) propagation time along the fiber as a function of the glass index of refraction and wavelength. The term in the brackets represent the group index of refraction which is given as:

$$n_{\mathrm{g}} = n - \lambda\frac{\mathrm{d}n}{\mathrm{d}\lambda} . \tag{7.71}$$

Bearing in mind that according to the Sellmeier equation the index of refraction decreases with the wavelength we have $\mathrm{d}n/\mathrm{d}\lambda < 0$ and thus $n_{\mathrm{g}} > n$. According to equation (7.70) for the scatter of arrival times at the fiber end the following is obtained:

$$\text{ä}\tau = \frac{\mathrm{d}\tau}{\mathrm{d}\lambda}\text{ä}\lambda = \frac{L}{c}\text{ä}\lambda\frac{\mathrm{d}}{\mathrm{d}\lambda}\left(n - \lambda\frac{\mathrm{d}n}{\mathrm{d}\lambda}\right)$$
$$= -\frac{L}{c}\lambda\frac{\mathrm{d}^2 n}{\mathrm{d}\lambda^2}\text{ä}\lambda \quad . \tag{7.72}$$

The corresponding contribution of the material dispersion D_M to the dispersion coefficient is given by:

$$D_{\mathrm{M}} = \frac{1}{L}\frac{\mathrm{d}\tau}{\mathrm{d}\lambda} = -\frac{\lambda}{c}\frac{\mathrm{d}^2 n}{\mathrm{d}\lambda^2}\,. \qquad (7.73)$$

Sometimes, especially in the cases when the signal occupies a broad spectral range and when the dispersion variation on the wavelength must be taken into account, it is of interests to introduce the dispersion slope as:

$$S_{\mathrm{M}} = \frac{\mathrm{d}D_{\mathrm{M}}}{\mathrm{d}\lambda}\,. \qquad (7.74)$$

Besides the material dispersion there is also waveguide dispersion in the fibers. For the step index fiber the waveguide dispersion the following is valid:

$$D_{\mathrm{W}} = -\frac{n_2 \Delta}{c\lambda} V \frac{\mathrm{d}^2}{\mathrm{d}V^2}(bV)\,, \qquad (7.75)$$

where we have:

$$b = \frac{\beta^2 - k_1^2}{k_2^2 - k_1^2}\,, \qquad (7.76)$$

which is given without explicitly deriving. In the case where V number takes large values the dimensionless quantity b tends to unity where at the cutoff of each mode there is $b = 0$ is valid.

The profile dispersion occurs when one takes into account that the normalized refractive index difference is not constant but depends slightly on wavelength. This contribution is typically small and can be neglected. In figure 7.12. it is presented the typical overall chromatic dispersion ($D = D_M + D_W + D_P$) dependence on the wavelength.

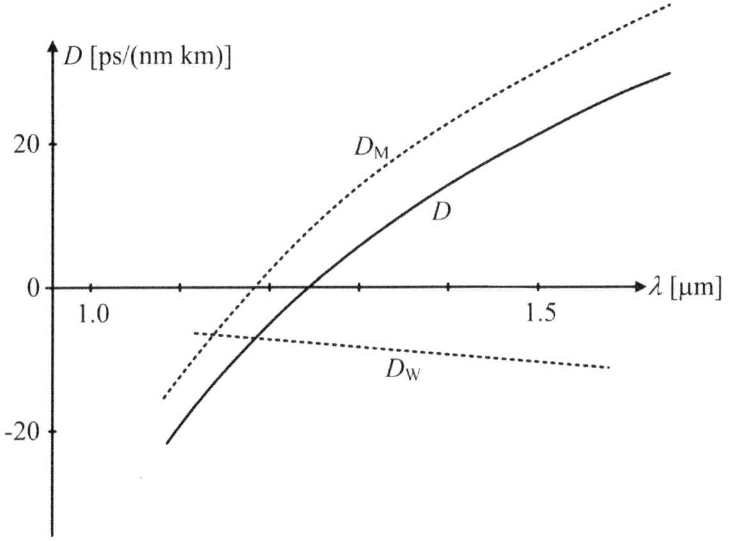

Figure 7.12. Total dispersion as a result from the material contribution of the core and from the waveguide contribution.

Further we will analyze the impact of the chromatic dispersion on the propagation of a light pulse, which will be considered as a monochromatic wave multiplied, i.e. modulated by a corresponding envelope function. Typically, a Gaussian function has been chosen for the envelope function. Therefore, the temporal profile of the pulse irradiance is:

$$I(t) = I_0 \exp\left[-\left(\frac{t}{T}\right)^2\right], \qquad (7.77)$$

where I_0 is the peak irradiance and T the pulse duration. Sometimes as a pulse duration the full width at half maximum (FWHM) has been taken into account. In the case of a Gaussian shaped pulse the relation between the FWHM τ and the pulse duration T is given by:

$$\tau = 2\sqrt{h\ 2T}. \qquad (7.78)$$

When a Gaussian shaped pulse propagates along the fiber length L both pulse parameters duration and peak irradiance are modified. It is not difficult to obtain the following relation for the pulse duration:

$$T_L = T\sqrt{1+\left(\frac{L}{l_D}\right)^2}, \qquad (7.79)$$

where for the dispersion length l_D is valid:

$$l_D = \frac{T^2}{|\beta_2|}, \qquad (7.80)$$

and where the β_2 parameter is obtained as:

$$\beta_2 = \left.\frac{d^2\beta}{d\omega^2}\right|_{\omega=\omega_0}, \qquad (7.81)$$

i.e. as the second derivative of the propagation constant $\beta(\omega) = \omega n(\omega)/c$ for the given central angular frequency ω_0. According to the equation (7.79) after passing the fiber length, which is equal to the dispersion length the pulse, the pulse duration is increased by a factor $\sqrt{2}$. In the case of a very long fiber length, i.e. when $L > l_D$ is satisfied, one obtain the following pulse duration at the fiber end:

$$T_L = \frac{|\beta_2| L}{T}, \qquad (7.82)$$

where an interesting situation is obtained where the shorter the pulse duration at one end of the fiber the longer the pulse duration at the other fiber end.

Polarization Mode Dispersion

Polarization mode dispersion is a result of the pulse broadening due to polarization of a fiber propagating optical pulse and the birefringence of the fiber. An optical pulse that travels along the fiber may consist of different polarization states where each state corresponds to a slightly different index of refraction. There are many reasons for different indices of refraction for different polarization states such as mechanical stress in the fiber, fiber bends, variations in the fiber structure, and the deviation from the circularity of the fiber. Therefore, the different polarization states experience different indices of refraction thus propagating with different velocities, causing pulse broadening and thus polarization mode dispersion (PMD). In figure 7.13. the illustration of the polarization mode dispersion is presented.

We can consider that an optical pulse consists of two orthogonal polarizations. Due to the fiber birefringence two modes travel at different group velocities. Therefore, they start to separate with the pulse propagation along the fiber. It should be pointed out that unlike other dispersion phenomena, the polarization mode dispersion has a random nature. The reason for this lies in the fact that PMD causes many different physical phenomena, which generally have random nature such as stress and temperature. However, PMD effects are small and become important at very high data rates (10 Gbps and above). As an effect of the PMD random nature, the delay between two orthogonal modes accumulates along the distance, but in opposite with other deterministic dispersions where the time effects are linearly added along the fiber, in the case of PMD the addition of the time delay of each elementary fiber length is random and thus the total time delay is proportional to the square root of the fiber length:

$$\tau_{PMD} = D_{PMD}\sqrt{L}, \quad (7.83)$$

where D_{PMD} is the polarization dispersion coefficient, given in ps km$^{-1/2}$, and L is the fiber length. The polarization dispersion coefficient is typically a function of the fiber construction and has the value somewhere in the range of 0.1 – 1 ps km$^{-1/2}$.

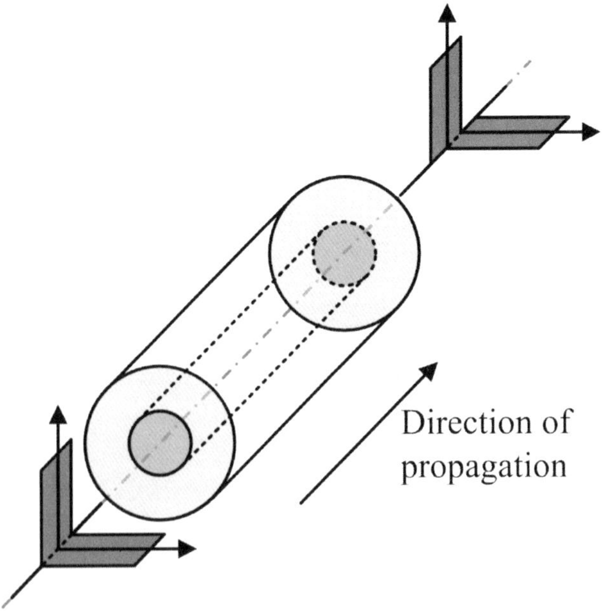

Figure 7.13. Polarization mode dispersion as a result of fiber birefringence.

The polarization state of the propagating wave cannot be maintained along the standard fiber. Therefore, in order to maintain the polarization state one can try to reduce the birefringence. However, this can be hard to achieve, even if the built-in tension can be suppressed during the manufacturing process, there is still no control over bending the fiber by the users. As it is not possible to suppress the birefringence to the negligible levels, one can try to do quite opposite by making the fibers intentionally with large birefringence. One can make such fibers by introducing an asymmetric core structures such as elliptic core or insert additional structural elements that break the fiber circular symmetry. By inserting elements with a slightly different thermal expansion coefficient, where during the cooling of the glass of the fiber a mechanical stress has been built into the fiber, the symmetry of the fiber is broken. Typical structures of such polarization maintaining fibers are presented in figure 7.14.

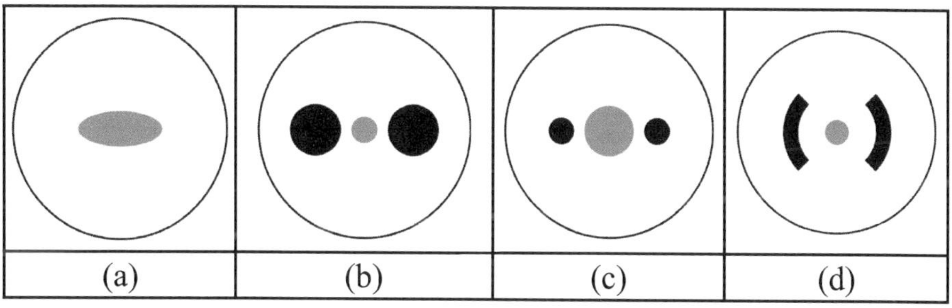

Figure 7.14. Polarization-maintaining fiber structures: (a) elliptical core, (b) panda, (c) pits, and (d) bowtie fiber.

Fiber Losses

As mentioned at the beginning of this Chapter there was a long road in reducing the fiber loses from the former value of 20 dB/km to the today's typical value of 0.2 dB/km, which represents the fundamental limit. This limit is determined by three factors:

- The material resonances in the ultraviolet,
- The material resonances in the infrared, and
- The Rayleigh scattering as a result of the glass statistical structure.

It is rather difficult task to theoretically determine the lowest possible loss of an optical fiber. By considering the bulk glass data one can derive that the minimal loss is 0.114 dB/km which corresponds to the energy loss of 2.6%/km. However, this theoretical level hasn't achieved yet, but the loss of 0.2 dB/km has been routinely obtained since the late 1980s.

In figure 7.15. is presented the wavelength dependence of a standard optical fiber loss. One can notice that due to the impurity molecules in the glass there is a local loss peaks between the second and third windows at about at 0.95, 1.25 and 1.38 μm. This loss peaks are caused by the water molecules vibrations which can be found embedded in the glass. Besides the water there are other impurities which can contribute the overall loss. Typically, among these impurities are metals Fe, Cu, Co, Cr, Ni, and Mn. For example, the copper (Cu) concentration of only 10^{-9} can cause loss of several tenths of dB/km.

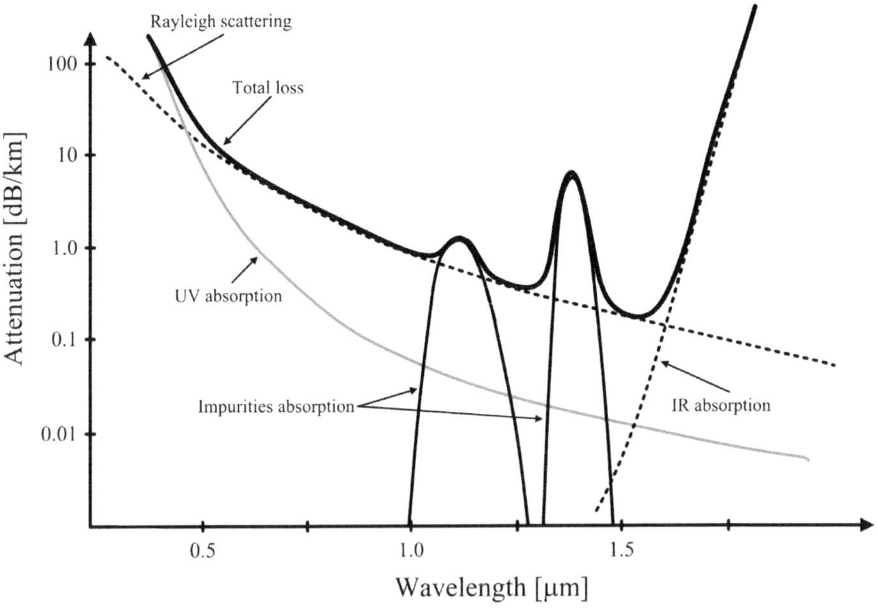

Figure 7.15. Wavelength dependence of a standard optical fiber loss.

As can be seen from figure 7.15. infrared (IR) absorption tail causes attenuation for the wavelengths longer than 1.6 μm. The lowest loss of about 0.18 dB/km is obtained for the wavelength region of 1.55 μm for the standard optical fiber. This loss is very close to the fundamental scattering limit.

FIBER-OPTIC COMMUNICATIONS SYSTEMS

Any fiber-optic communications system consists at least of three basic elements: optical fiber, optical transmitter (source), and optical detector (receiver). Therefore, their limiting parameters will determine the characteristics of any fiber-optic communications system, such as highest possible data rate, longest possible optical links, highest possible coverage, etc. Depending on the service, which can be offered based on the given the fiber-optic communications system, we can distinguish different network topologies. Therefore, from the network topology point of view there are three basic categories: point-to-point links, distribution networks, and local-area networks, which will be in the focus in this chapter.

Point-to-Point Links

The simplest possible lightwave communications systems are certainly point-to-point links. Their main role is to transport information from one point to another, where the information, which is in the form of a digital bit stream, must be transported as accurately as possible. The length of such link varies from less than a kilometer, i.e. the short haul links to thousands of kilometers, i.e. the long haul links, which depends on the links actual application. As example for a short haul links we can use an optical data links used to connect the computers and terminals within the same building or between two relatively adjacent buildings. Such links do not require low loss and/or wide bandwidth of optical fibers. For a long haul links, e.g. undersea lightwave systems, the use of low losses and a large bandwidth of optical fibers are of paramount importance where the reducing the overall operating cost is one of the main issues.

If the data links lengths exceed a certain value, typically in the range of 20 – 100 km, which depends on the transmitter wavelength, it is necessary to compensate the fiber losses. Otherwise, the signal becomes too weak to be detected with a certain given probability. In order to compensate the fiber losses there are two schemes which are commonly used that are presented in figure 7.16. At the early stages of the optical links development the regenerators are exclusively used for the loss compensations. The regenerator consists of the receiver-transmitter pair, which detects the incoming optical signal, recovers the electrical bit stream, and finally converts it back into optical form. Since the invention of the optical amplifiers, which has happened around 1990, fiber losses are compensated by using optical amplifiers. Optical amplifier amplifies the optical bit stream directly without converting it in the electric domain. Optical amplifiers are especially useful for the wavelength-division multiplexed (WDM) lightwave systems where they can manage the amplification of many channels simultaneously.

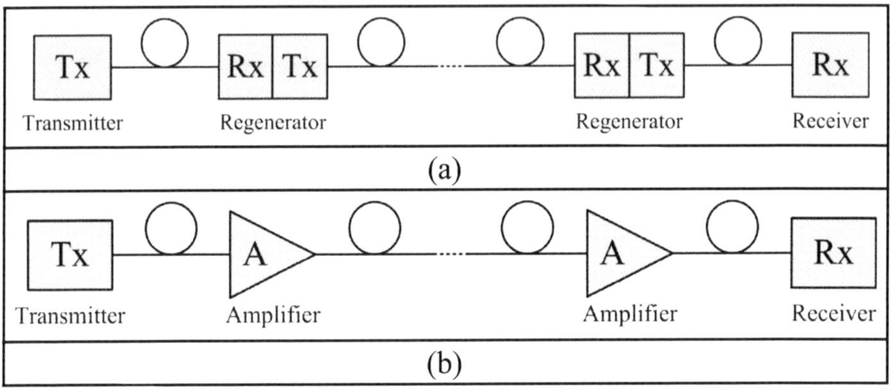

Figure 7.16. The compensation of point-to-point fiber links losses with the help of (a) regenerators and (b) optical amplifiers.

Distribution Networks

When we have a large number of customers to whom the information must not only be transmitted but as well distributed the point-to-point network topology cannot provide such service. Typical examples are local-loop distribution of the telephone services and the broadcast of the video channels over cable television (CATV - ***Common-Antenna TeleVision***). There are two topologies for the distribution networks which are presented in figure 7.17. For the hub topology, the corresponding channel distribution occurs at the central locations, which are commonly named hubs. This type of network is commonly called metropolitan-area networks (MANs) as the hubs are typically located in the major cities. In this case fiber role is similar as in the case of point-to-point networks. Telephone networks use hub topology for distribution of voice channels within a city.

Bus topology is based on a single fiber cable that carries the multichannel optical signal. The distribution is performed by using an optical tap, which redirect a small portion of the power of the optical signal to each subscriber. Typically, CATV systems use bus topology. The use of an optical fiber allows distribution of a large number of channels due to its large bandwidth compared with coaxial cables. Moreover, the distribution of the high-definition television (HDTV) channels requires the lightwave transmission due to a large bandwidth of each video channel. The main problem of the bus topology is that the signal loss increases exponentially with the number of used taps and thus limits the number of subscribers connected to the same optical bus.

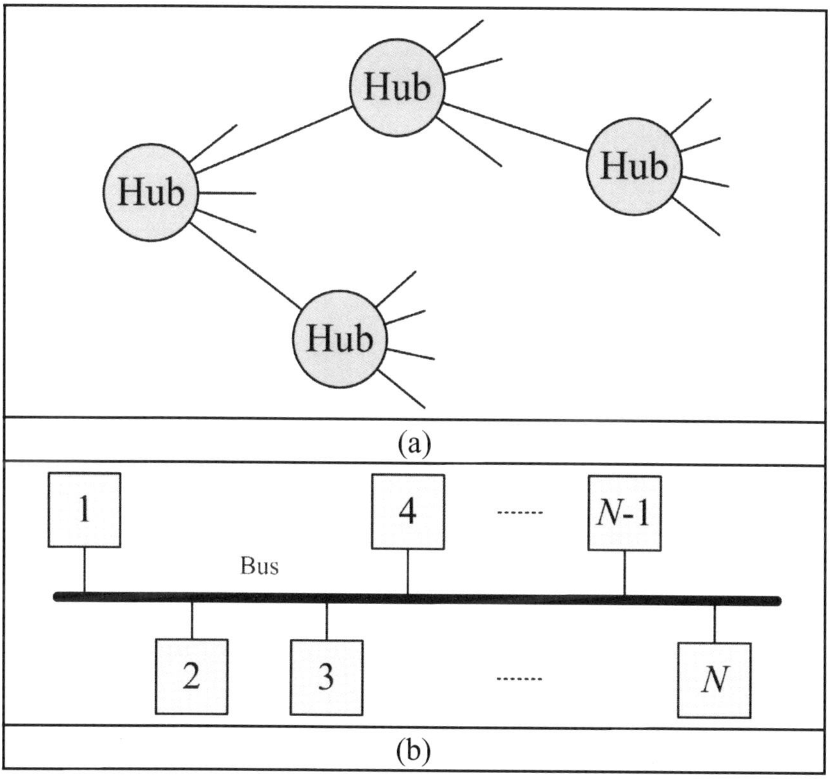

Figure 7.17. Distribution networks with (a) hub topology and (b) bus topology.

Local Area Networks

In the cases when there is a large number of users within a relatively small area, e.g. a university campus or a large company facility, etc. where the users are interconnected in a way that each user can access the network randomly to transmit data to any other user we have a so called local area network (LAN). The transmission distances in the optical access networks of a local subscriber loop are relatively short (less then 10 km), so the fiber losses are not an issue for LAN applications. The advantage of using optical access over other techniques is the large bandwidth of the fiber-optic based communication systems.

There are two main topologies used for the LAN applications the ring and the star topology which are presented in figure 7.18. The adjacent nodes are connected by point-to-point links to form a closed ring in the ring topology case. Each node transmits and receives the data where the corresponding transmitter-receiver pair acts also as a repeater. The fiber-optic LANs based on the ring topology has standardized interface known as the fiber distributed data interface (FDDI), which operates at 100 Mb/s (data rate) by using the multimode fibers and the transmitters based on the 1.3 μm light-emitting diodes (LEDs).

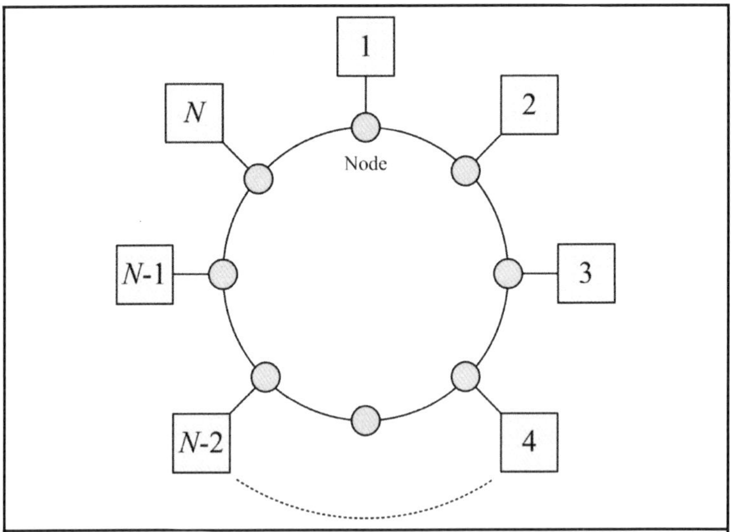

(a)

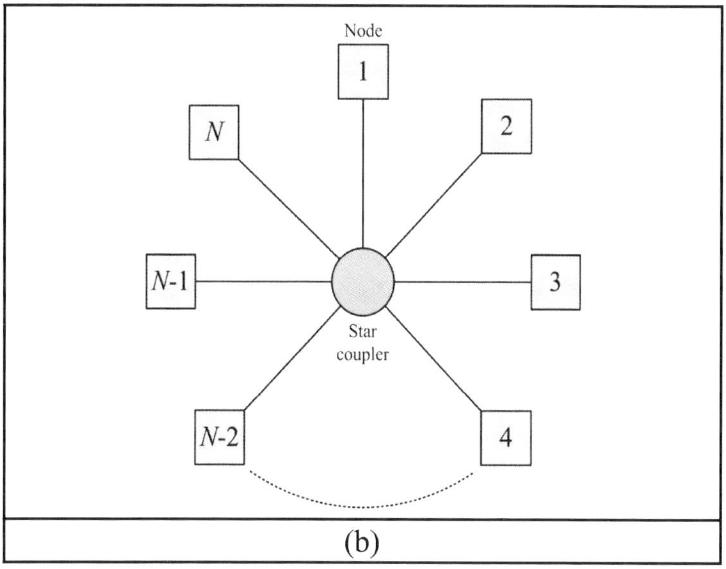

Figure 7.18. Local-area networks with (a) ring topology and (b) star topology.

Regarding the star topology all nodes are connected with the help of the point-to-point links to a central node called a hub or a star. Depending on whether the central node is an active or passive device the star topology based LANs can be further divided into two groups: active-star and passive-star networks. The active-star configuration converts all incoming optical signals into the electrical domain through optical receivers. The corresponding electrical signal is then distributed to the node transmitters. In the case of the passive-star configuration, the distribution has been performed place in the optical domain through the directional couplers.

Fiber-Optic Network Design Consideration

In order to design a fiber-optic communication system one requires a clear insight in the limitations which are imposed by the fiber loss, dispersion and nonlinear effects. As the wavelength affects the fiber properties the right choice of the operating wavelength is a major design issue. The two most important characteristics of any fiber-optic communication system are the bit rate and the transmission distance. Therefore, here it will be discussed how the fiber loss and dispersion affect the data rate and transmission distance.

Fiber losses play an important role in the system design except in the case of some short-haul fiber links. Let P_T be the optical power launched by the transmitter in the optical fiber. In order to decipher the data stream with the bit rate B the receiver at the other end of the fiber-optic link requires a minimum average power P_R. The corresponding maximal fiber length between the transmitter and the receiver is given by:

$$L = \frac{1}{\alpha} 10 \log_{10}\left(\frac{P_T}{P_R}\right), \qquad (7.84)$$

where α f is the fiber cable net loss (in dB/km) including splice and connector losses. Bearing in mind that $P_R = Nh\nu B$ where $h\nu$ is the corresponding photon energy and N is the average number of photons per bit required by the receiver in order to achieve the minimum bit error rate (BER), the length L decreases logarithmically as B increases at a given operating wavelength. If the transmitted power is taken to be $P_T = 1$ mW at $\lambda = 1.55$ μm, the minimum required number of photons per bit is $N = 10$ for a bit error rate of $BER = 10^{-9}$, so the maximal distance than can be reached with such a system is $L_{max} = 245$ km where the bit rate of $B = 10$ Gbps and fiber cable loss of $\alpha = 0.2$ dB/km was assumed.

In order to ensure that enough power will reach the receiver and provide reliable signal detection, a proper power budget of an optical link must be calculated during the design process. The receiver sensitivity represents the minimum average power required by the receiver, which must be provided by the transmitter with the given emitting power. The power budget is much simpler to present in decibel units with optical powers expressed in dBm units thus we have:

$$P_T = P_R + C_L + M, \qquad (7.85)$$

where C_L is the total channel loss and M is the system margin. With the system aging the component parameters may degrade during their lifetime or other unforeseen events thus reaching the low levels of the receiver captured power, which can be even below its sensitivity. Therefore, the system margin introduces a certain reserve in the design where the parameter degradation cannot affect the receiver sensitivity. Typical value of the system margin that has been taken during the system design is 4 – 6 dB.

The channel loss C_L includes all possible losses along the link such as connector and splice losses, which can be written as following:

$$C_L = \alpha L + \alpha_C N_C + \alpha_S N_S, \qquad (7.86)$$

where α_C and α_S are the connector and splice losses, respectively, and N_C and N_S are the connector and splice number along the link, respectively. Sometimes the splice loss is included with the specified fiber cable loss. The connector loss must include the connectors at the transmitter and receiver ends as well as other connectors if used within the fiber link.

Typically connectors for a multimode fiber have a losses in the range of 0.2 – 0.5 dB per connector, whereas the connectors for the single-mode fiber, which are factory made and fusion spliced, will have losses of 0.1 – 0.2 dB. Field terminated single-mode connectors may have higher losses somewhere in the range of 0.5 – 1.0 dB.

Splicing the multimode fibers are typically performed with the mechanical splices, although there is a possibility of fusion splicing as well. Due to the large core fusion splicing has about the same loss as mechanical splicing. However, the fusion is more reliable in adverse environments. Typical value of multimode splice losses is in the range 0.1 – 0.5 dB. Fusion splicing of a single-mode fiber typically has less than 0.05 dB.

In the case of a step index multimode fibers equation (7.14) provides an approximation for the upper limit for the product of the maximal bit rate B and the fiber length L. According to the equation (7.14), where the relation for the difference in time travel is given, we have:

$$B = \frac{c}{n_2 \Delta}, \qquad (7.87)$$

where for the values $n_2 = 1.46$ and $\Delta = 0.01$ even at a low bit rate of only 1 Mbps we have that such multimode fiber-optic systems are dispersion-limited where their transmission distance is limited to below than 5 km. Therefore, the multimode step index fibers are very rarely used for designing the fiber-optic communication systems. Much better situation regarding the dispersion-limited fiber-optic communications system is when gradient optical fibers are used. In the case of a gradient multimode fibers equation (7.17) provides an approximation for the upper limit for the product of the maximal bit rate B and the fiber length L. According to the equation (7.17), where the relation for the difference in time travel is given, we have:

$$B = \frac{2c}{n_2 \Delta^2}. \qquad (7.88)$$

By comparing the equations (7.87) and (7.88) one have considerable improvement in the product of the maximal bit rate and the fiber length of two orders of magnitude if $\Delta = 0.01$ in the case of gradient fibers. The first generation of fiber-optic telecommunication systems took advantage of such an improvement and used graded index fibers where the first commercial system operated at a bit rate of 45 Mbps with a repeater spacing of less than 10 km.

The second generation of the fiber-optic communication systems uses mainly single-mode fibers near the minimal dispersion wavelength of about 1.31 μm. The main limiting factor in these systems is the pulse broadening induced by dispersion due to the large source spectral width. For the product of the maximal bit rate and the fiber length we have:

$$B = \frac{1}{4|D|\Delta\lambda}, \qquad (7.89)$$

where $\Delta\lambda$ is the source spectral width and D is the total dispersion of the fiber. The value of $|D|$ mainly depends on how close we are to the zero-dispersion wavelength of the fiber and is typically ~ 1 ps/(km nm). For the spectral width of the source of $\Delta\lambda = 2$ nm we have $|D|\Delta\lambda$ ~ 2 ps/km, so finally is obtained BL = 125 Gbps km. Therefore, the second generation systems are mainly loss-limited up to the bit rates of about 1 Gbps.

The third and fourth generations of lightwave systems operate near 1.55 μm where the minimal fiber loss occurs. In this spectrum range fiber dispersion becomes a major limitation as D ~ 16 ps/(km nm) for the standard silica fibers. The relation which gives the limit for the corresponding bit rate and the fiber length of such a communication system is given by:

$$B^2 L = \frac{1}{16\,|\beta_2|}, \qquad (7.90)$$

where β_2 is the second derivative of the propagation constant. Typically, the 1.55 μm operating ligthwave systems become dispersion-limited only for $B > 5$ Gbps.

Coherent Fiber-Optic Communication Systems

Coherent detection is based on the combining of the received optical signal coherently with a continuous-wave (CW) optical field, which is generated by the local oscillator LO, before it impinges the photodetector D as presented in figure 7.19. At the receiver R output a corresponding electrical signal is obtained. Local oscillator generates a CW field by using a narrow linewidth laser.

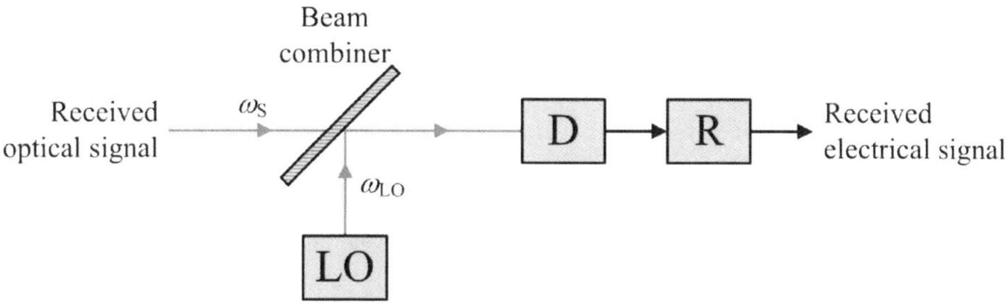

Figure 7.19. Coherent detection of an optical signal.

In order to gain an insight how the mixing of the received optical signal with the local oscillator output can improve the receiver performance, we will start from the following optical signals in complex domain:

$$E_S = E_{S0}\exp[-j(\omega_S t + \varphi_S)], \quad (7.91)$$

where ω_S is the carrier frequency, E_{S0} is the amplitude, and φ_S is the phase. The optical field associated with the local oscillator is given similarly as:

$$E_{LO} = E_{LO0}\exp[-j(\omega_{LO} t + \varphi_{LO})], \quad (7.92)$$

where E_{LO0}, ω_{LO}, and φ_{LO} are the corresponding amplitude, frequency, and phase of the local oscillator, respectively. We will further assume that both fields E_S and E_{LO} have identical polarization. As the photodetectors are sensitive to the captured optical power, the optical power incident at the photodetector is given by:

$$P(t) = K|E_S + E_{LO}|^2, \quad (7.93)$$

where K is a proportionality constant. Equations (7.91), (7.92), and (7.93) further gives:

$$P(t) = P_S + P_D + 2\sqrt{P_S P_D}\cos(\omega_F t + \varphi_S - \varphi_D), \quad (7.94)$$

where we have:

$$P_S = K|E_S|^2, \quad P_D = K|E_D|^2, \quad \omega_F = \omega_S - \omega_D, \quad (7.95)$$

where ω_{IF} is the intermediate angular frequency. In the case when $\omega_S \neq \omega_{LO}$ is fulfilled the optical signal is demodulated in two stages. In the first stage the carrier frequency is converted to an intermediate frequency $\nu_{IF} = \omega_{IF}/(2\pi)$, typically in the range of 0.1 – 5 GHz, before the signal is demodulated to the baseband. Depending on the intermediate frequency value there are two different coherent detection techniques. If the intermediate frequency is equal to zero we have a homodyne detection technique and if not the heterodyne detection technique.

The homodyne detection technique has the local oscillator frequency identical with the corresponding signal carrier frequency, so we have $\omega_{IF} = 0$. In this case equation (7.94) becomes:

$$P(t) = P_S + P_D + 2\sqrt{P_D P_S}\cos(\varphi_S - \varphi_D), \quad (7.96)$$

where for the electrical current of the photodetector we have:

$$\begin{aligned} I(t) &= \Re P(t) \\ &= \Re\left[P_S + P_D + 2\sqrt{P_D P_S}\cos(\varphi_S - \varphi_D)\right], \end{aligned} \quad (7.97)$$

where $\Re$ is the detector responsivity. Typically, it is fulfilled $P_{LO} >> P_S$, and $P_S + P_{LO} \approx P_{LO}$, where the last term in equation (7.97) carries the transmitted information. If the local oscillator phase is locked to the signal phase $\varphi_S = \varphi_{LO}$, then the homodyne signal is given by:

$$I(t) = 2\Re\sqrt{P_D P_S}. \quad (7.98)$$

If compared with the direct detection of an optical signal, where the photodetector current is given by $I_D(t) = \Re P_S(t)$, the homodyne detection provides an electrical current which power is increased by a factor $4P_D/\langle P_S(t)\rangle$, where $\langle P_S(t)\rangle$ denotes the received optical signal average power. As the local oscillator power is much larger than the received optical signal power ($P_{LO} >> P_S$) the power enhancement can exceed

20 dB. Although the corresponding detector shot noise is also increased, the signal-to-noise ratio (SNR) is enhanced by a large factor.

One of the advantages of the coherent detection is hidden in the last term of equation (7.97) where the phase dependence of the photodiode current can be used in order to transmit the phase modulated signal. A disadvantage of the homodyne technique also originates from the phase sensitivity. As the phase of the local oscillator fluctuates it can influence the overall system noise. However, the signal and local oscillator phase difference can be kept constant through an optical phase-locked loop (PLL) thus eliminating the corresponding local oscillator phase noise. It is worth to mention that the implementation of an optical phase-locked loop is not a simple task.

Heterodyne detection technique uses the local oscillator with the frequency which differs from the signal carrier frequency such that the intermediate frequency is in the microwave region ($\nu_{IF} \sim 1$ GHz). The photocurrent in the case of a heterodyne detection is given by:

$$I(t) = \Re(P_S + P_D) + 2\Re\sqrt{P_D P_S}\cos(\omega_F t + \varphi_S - \varphi_D) \quad (7.99)$$

As in the case of the homodyne detection we also have the following conditions fulfilled: $P_{LO} >> P_S$, and $P_S + P_{LO} \approx P_{LO}$. The DC component, which is contained in the first term of equation (7.99), can be easily removed with the help of a simple band-pass filter. In this case the AC component of the photocurrent is given with:

$$I_{AC}(t) = 2\Re\sqrt{P_D P_S}\cos(\omega_F t + \varphi_S - \varphi_D) \quad (7.100)$$

Information can be transmitted as in the case of the homodyne detection through the amplitude, phase, or frequency modulation of the optical carrier. The local oscillator amplifies the received signal by a large factor thus increasing the signal-to-noise ratio. However, the signal-to-noise ratio improvement is lower by a factor of approximately 3 dB if compared with the homodyne detection. The origin of the reduction in signal-to-noise ratio can be found in equations (7.98) and (7.100) where the average power of the heterodyne photocurrent is reduced by a factor 2 when $I_{AC}^2(t)$ is averaged over a full cycle at the intermediate frequency.

Although there is a 3 dB penalty the receiver design is much simpler as there is no need for the phase-locked loop. The fluctuations of the carrier and local oscillator phases should be controlled by using narrow linewidth semiconductor lasers for both sources. Therefore, this feature makes the heterodyne detection technique quite suitable for practical implementation in the coherent fiber-optic communication systems.

Considerable signal-to-noise enhancement is one of the most prominent advantages of the coherent detection techniques. In order determine the signal-to-noise improvement an analysis of a noise of such a system will be performed. There are two main noise sources of a receiver the shot noise and the thermal noise. The corresponding variance of the photocurrent fluctuation is given by:

$$\left\langle i_n^2 \right\rangle = \left\langle i_S^2 \right\rangle + \left\langle i_T^2 \right\rangle, \qquad (7.101)$$

where $\left\langle i_S^2 \right\rangle$ is the variance of the photocurrent fluctuation caused by the shot noise and $\left\langle i_T^2 \right\rangle$ is the variance of the photocurrent fluctuation caused by the thermal noise for which is valid:

$$\left\langle i_S^2 \right\rangle = 2q\left(I + I_D\right)BW$$
$$\left\langle i_T^2 \right\rangle = \frac{4k_B TF}{R_L} BW, \qquad (7.102)$$

where q is the proton charge, I is the photocurrent, I_D is the photodetector dark current, is BW is the receiver bandwidth, k_B is the Boltzmann constant, T is the receiver absolute temperature, F is the receiver noise factor, and R_L is the receiver load resistance. The corresponding signal-to-noise ratio is obtained by dividing the average signal power by the average noise power. Therefore, in the heterodyne detection case, we have the following:

$$SNR = \frac{\left\langle I_{AC}^2(t) \right\rangle}{\left\langle i_n^2 \right\rangle} = \frac{\Re^2 P_D \left\langle P_S(t) \right\rangle}{q\left(\Re P_D + I_D\right)BW + \frac{2k_B TF}{R_L} BW}. \qquad (7.103)$$

In the case of a homodyne detection the signal-to-noise ratio is larger by a factor of 2 if $\varphi_S = \varphi_{LO}$ is assumed. The power of the local oscillator power can be easily controlled and it can be made large enough that the receiver noise is dominated by shot noise, or $\langle i_{S}^{2} \rangle > \langle i_{T}^{2} \rangle$, so we have:

$$P_D > \frac{2k_B F}{q\Re R_L} \quad (7.104)$$

If the dark current contribution is negligible ($I_D < \Re P_D$) the signal-to-noise ratio is then given by:

$$SNR \approx \frac{\Re\langle P_S(t)\rangle}{qBW} \quad (7.105)$$

FIBER-OPTIC SENSORS

In parallel with the fiber-optic communication technology the fiber-optic sensor technology becomes the major user of technology associated with the optoelectronic and fiber-optic communications industries. Moreover, many of these components were often developed for fiber-optic sensor applications. Along with the price reduction and the quality improvements of the optoelectronic and fiber-optic components the possibility of substituting the traditional sensors for various physical quantities such as temperature, pressure, vibration, linear and angular position, strain, humidity, concentrations, etc. for sensor applications has been enhanced. The advantage of the fiber-optic sensors over classical sensors one can see in their ability to be lightweight, of very small size, passive, resistant to electromagnetic interference, high sensitive, etc. In the following section the basic types of fiber-optic sensors will be reviewed and followed by the description of how these sensors can be applied.

Basic Concept of Fiber-Optic Sensors

Typically, fiber-optic sensors are roughly grouped into two basic classes usually called the extrinsic, or hybrid, fiber-optic sensors and intrinsic, or all-fiber, sensors. The simplified model of an extrinsic fiber-optic sensor is presented in figure 7.20. The input fiber leads the light into the light modulator which imprints the information onto the propagating light beam as a response of environmental effects. The information can be stored in the intensity, phase, frequency, polarization, spectral content, etc. of a propagating wave. Finally, the output optical fiber further carries the impressed information toward an optoelectronic processing unit where the information can be extracted.

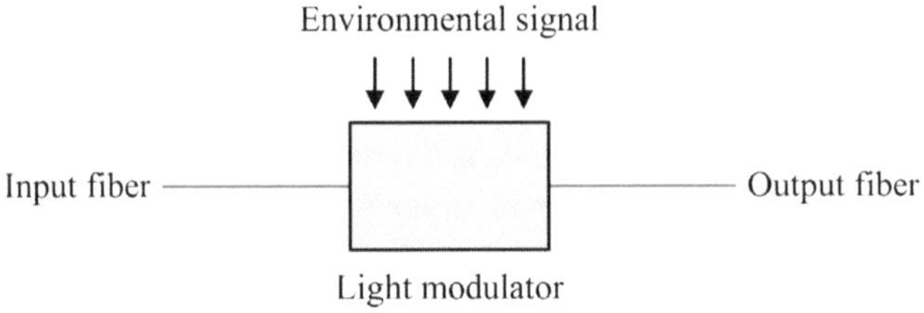

Figure 7.20. Extrinsic-fiber optic sensor.

If the input optical fiber also acts as the output fiber and where the environmental effect impresses information onto the light beam while it is in the fiber we have an intrinsic or an all-fiber sensor which is presented in figure 7.21.

In its simplest possible form, the fiber-optic sensor is usually based on the intensity modulation. For example, a simple intensity modulated fiber-optic sensor aimed for the vibration measurement is presented in figure 7.22. where a simple closure of the vibration sensor consists of two optical fibers held in close proximity to each other. The light, which input fiber emits in the free space, expands into a cone of light whose angle depends on the numerical aperture of the fiber. The amount of light, which is captured by the output fiber, depends on its acceptance angle (numerical aperture) and the distance *d* between the optical fibers. When the distance between them has been changed (due to the vibration for example) the power transmitted through the output fiber has been changed in the same manner thus giving the rise of the intensity modulation.

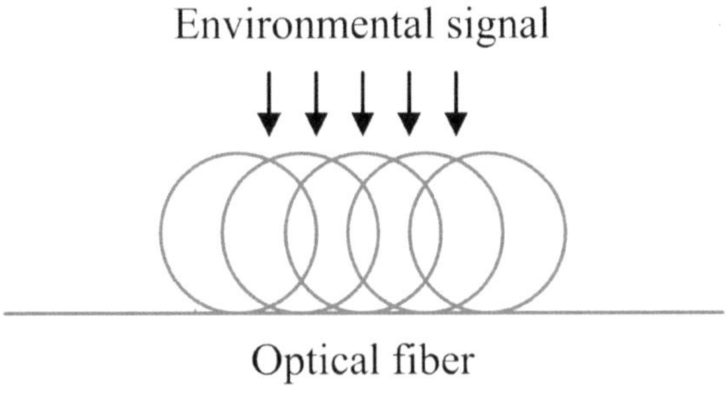

Figure 7.21. Intrinsic fiber-optic sensor.

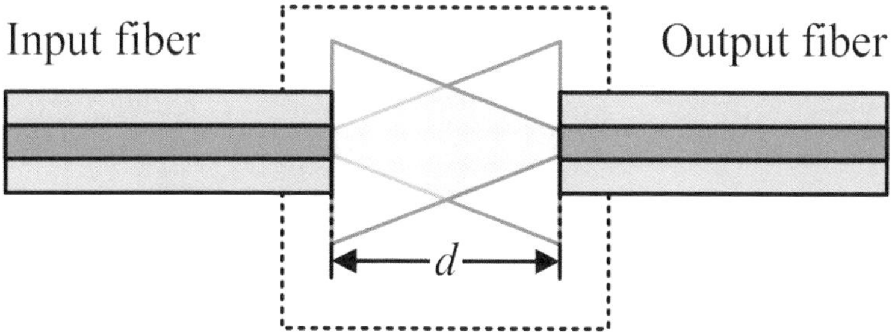

Figure 7.22. Simple fiber-optic sensor for vibration measurement.

Similar solution to the previous one is presented in figure 7.23. Instead of an in line fibers here a mirror has been flexibly mounted to the object which vibration is to be measured. As the mirror vibrates along with the object the effective separation between the optical fibers shifts modulates the captured intensity.

In order to measure lateral displacement a simple translation sensor can be configured as in presented in figure 7.24. The input fiber, which has been illuminated with the help of a light source LS illuminates further two output fibers. The light form the output fibers have been captured with the photodetectors PD1 and PD2. The photocurrents difference is proportional to the small translation displacements thus we have:

$$i_{\mathrm{PD1}} - i_{\mathrm{PD2}} \propto d, \quad (7.106)$$

where i_{PD1} and i_{PD2} are the photocurrents and d is the lateral displacement.

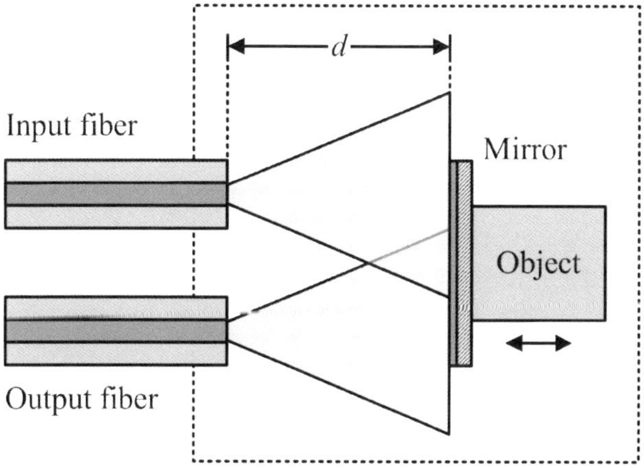

Figure 7.23. Simple fiber-optic sensor for vibration measurement by using a mirror.

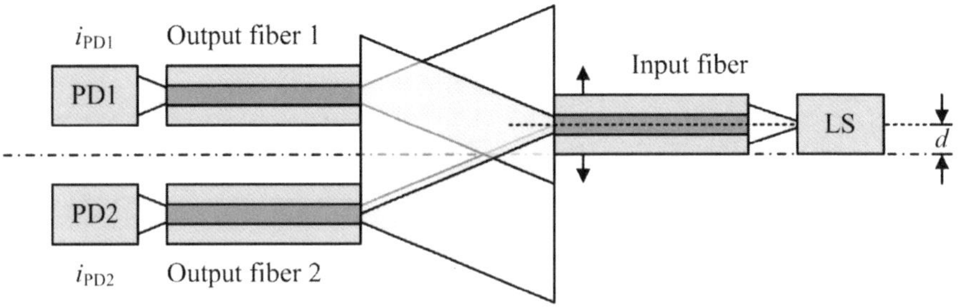

Figure 7.24. Simple fiber-optic translation sensor.

Interferometric Fiber-Optic Sensors

Based on different interferometer topologies such as Michelson, Mach-Zehnder, ring resonator, and Sagnac interferometers one can construct highly sensitive fiber-optic sensors aimed for measurement versatile physical quantities. Therefore, this section will be devoted to a brief review of different fiber-optic interferometer topologies for sensor applications and their main characteristics.

One of the main advantages of all-fiber interferometers, such as Michelson and Mach-Zehnder interferometers, is that their geometry can be extremely flexible but still keeping high sensitivity. Therefore, there is a possibility of constructing a wide variety of high-performance elements and fiber arrays, as presented in figure 7.25. where are presented the linear array, the planar array, the gradient element and the omnidirectional element.

In order to interrogate the phase change along the sensing fiber where the specifically designed sensing elements are employed, there is a variety of interferometric topologies. One of the most common is certainly the fiber-optic Mach-Zehnder interferometer which is presented in figure 7.26. As presented, the fiber-optic Mach-Zehnder interferometer consists of a light source, a coupler module, a transducer, and a homodyne demodulator. Usually, the light source is an isolated laser diode with a long coherence length. The light from the source is divided by the beam splitter, i.e. the coupler module in order to illuminate both arms of the interferometer. The transducer senses an environmental effect by exposing one of the interferometer arms to the measured parameter. In this way, the optical bath difference between the arms corresponds to the physical quantity to be measured. Finally, a homodyne demodulator is used to measure the optical path length difference and thus the corresponding physical quantity.

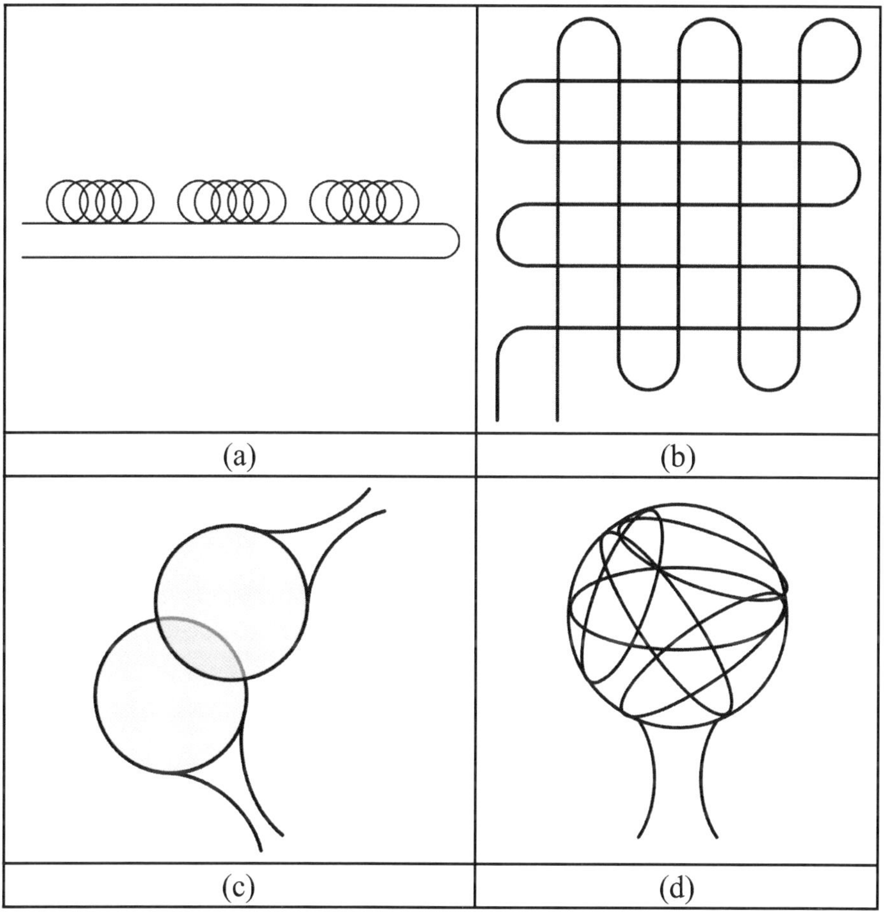

Figure 7.25. Flexible topology of interferometric fiber-optic sensors: (a) linear array, (b) planar array, (c) gradient element, and (d) omnidirectional element.

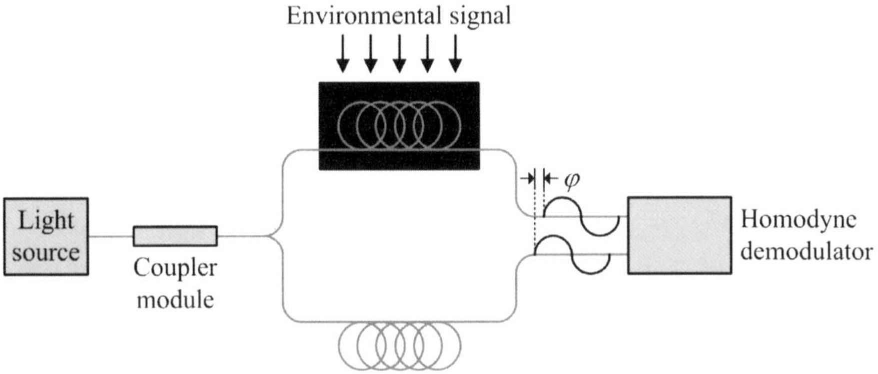

Figure 7.26. Fiber-optic Mach-Zehnder sensing interferometer basic elements.

The sensitivity of the Mach-Zehnder interferometers varies as a function of the relative phase of the light beams in two interferometer arms, as it is presented in figure 7.27. Due to other interfering environmental parameters such as temperature, pressure, humidity, etc., which can cause the additional phase shift and thus the signal fading, one can introduce, for example, a piezoelectric fiber stretcher into one of the arms and adjust the relative path length of the two arms in a way to gain the highest possible sensitivity. If the interferometer is set to the highest possible sensitivity it is usually said that the interferometer is in the quadrature.

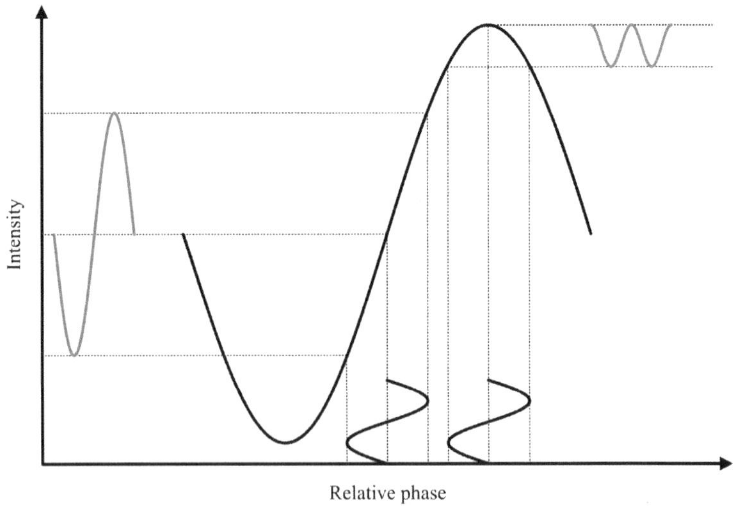

Figure 7.27. The sensitivity of the Mach-Zehnder interferometer.

The fiber-optic Michelson interferometer is in many respects similar to the Mach-Zehnder interferometer. In the case of the Michelson interferometer, instead of a homodyne demodulator, at the end of the interferometer arms the mirrors have been employed as presented in figure 7.28. As a result a very high level of the light is back reflected to the light source that can degrade the system performance. However, by using the diode pumped YAG (***Yttrium Aluminum Garnet***) ring lasers as light sources this problem can be overcame. Moreover, if the phase conjugate mirrors are used instead of a standard mirrors at the arms ends the polarization fading can be suppressed.

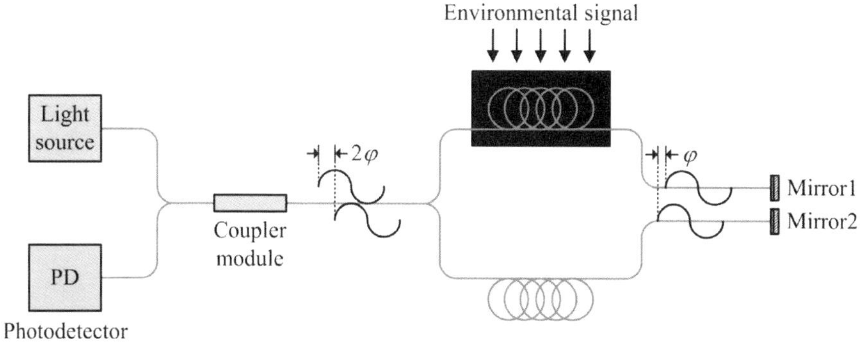

Figure 7.28. Fiber-optic Michelson sensing interferometer basic elements.

One of the most prominent characteristic of the fiber-optic sensors is their possibility to be multiplexed. Therefore, a large number of whether intrinsic or extrinsic sensor can be located along a single fiber-optic line. In other to address each sensor at the same fiber several techniques such as time, frequency, wavelength, coherence, polarization, and spatial multiplexing can be employed. An example of multiplexed fiber-optic sensors has been presented in figure 7.29. where the low-coherence light source (LCS) has been used in order to illuminate the sensors (S_k). This system uses the coherence multiplexing, where the corresponding back reflected light spectrum from each sensor has been interrogated with the help of an optical spectrometer (OS).

Typical physical quantities that can be measured with the multiplexed fiber-optic sensors are, for example temperature, pressure, strain, etc. The measured parameter typically modulates the cavity length of the intrinsic or extrinsic fiber-optic sensor thus producing different modulation patterns of the channeled spectrum. When the light from all the sensors has been captured by the optical spectrometer, each sensor can be extracted as it produces different modulation "frequency" of its channeled spectrum. Simply by applying the Fourier transform of the captured optical spectrum and finding the positions of the local peaks one can measure the corresponding cavity length and thus measure the physical quantity.

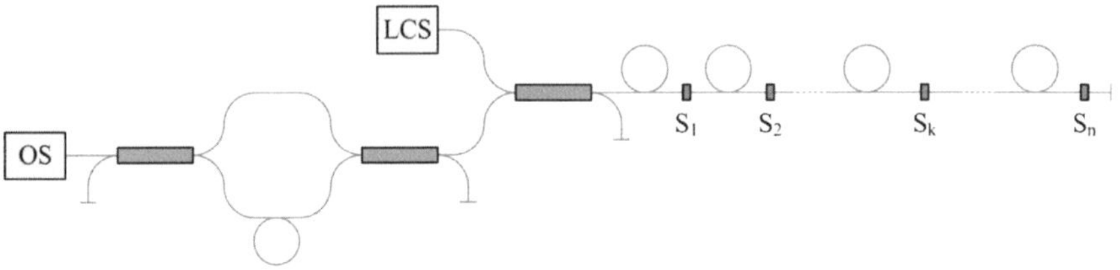

Figure 7.29. Multiplexing of fiber-optic sensors.

REFERENCES

1. John Powers. *An Introduction to Fiber Optic Systems (2nd Edition)*. New York: McGraw-Hill Professional, 1993.
2. John Senior. *Optical Fiber Communications*. London: Prentice-Hall International, Inc., 1985.
3. Ivan P. Kaminow and Tingye Li. *Optical Fiber Telecommunications IV A – Components*. San Diego: Elsevier Science (USA), 2002.
4. Casimer DeCusatis. *Handbook of Fiber Optic Data Communication (2nd Edition)*. San Diego: Academic Press, 2002.
5. Ivan P. Kaminow, Tingye Li, and Alan E. Willner. *Optical Fiber Telecommunications V A Components and Subsystems*. Amsterdam: Elsevier Inc., 2008.
6. Ivan P. Kaminow, Tingye Li, and Alan E. Willner. *Optical Fiber Telecommunications V B Systems and Networks*. Amsterdam: Elsevier Inc., 2008.
7. Cedric F. Lam. *Passive Optical Networks: Principles and Practice*. Amsterdam: Elsevier Inc., 2007.
8. Duwayne R. Anderson, Larry Johnson, and Florian G. Bell. *Troubleshooting Optical-Fiber Networks (2nd Edition)*. Amsterdam:

Elsevier, Inc., 2004.

9. Govind P. Agrawal. *Fiber-Optic Communications Systems (3rd Edition)*. New York: John Wiley & Sons, Inc., 2002.
10. Richard C. Dorf (Ed.). *Broadcasting and Optical Communication Technology*. Boca Raton: Taylor & Francis Group, LLC, 2006.
11. Casimer DeCusatis (Ed.). *Fiber Optic Data Communication: Technological Trends and Advances*. San Diego: Academic Press, 2002.
12. Jeff Hecht. *City of Light: The Story of Fiber Optics: The Story of Fiber Optics*. Oxford: Oxford University Press, Inc., 2004.
13. Avigdor Brillant. *Digital and Analog Fiber Optic Communications for CATV and FTTx Applications*. Bellingham: Society of Photo-Optical Instrumentation Engineers, 2008.
14. Enrico Forestieri (Ed.). *Optical Communication Theory and Techniques*. Boston: Springer Science + Business Media, Inc., 2005.
15. Keigo Iizuka. *Elements of Photonics Volume II For Fiber and Integrated Optics*. New York: John Wiley & Sons, Inc., 2002.
16. P. C. Becker, N. A. Olsson, and J. R. Simpson. *Erbium-Doped Fiber Amplifiers: Fundamentals and Technology*. San Diego: Academic Press, 1999.
17. Reinhold Noé. *Essentials of Modern Optical Fiber Communication*. Berlin: Springer-Verlag Berlin Heidelberg, 2010.
18. Barry Elliott and Mike Gilmore. *Fiber Optic Cabling (2nd Edition)*. Oxford: Newnes, 2002.
19. K. Thyagarajan and Ajoy Ghatak. *Fiber Optic Essentials*. Hoboken: John Wiley & Sons, Inc., 2007.
20. Casimer M. DeCusatis and Carolyn J. Sher DeCusatis. *Fiber Optic Essentials*. Amsterdam: Elsevier Inc., 2006.
21. Francis T. S. Yu and Shizhuo Yin (Eds.). *Fiber Optic Sensors*. New York: Marcel Dekker, Inc., 2002.
22. Fedor Mitschke. *Fiber Optics: Physics and Technology*. Heidelberg: Springer-Verlag Berlin Heidelberg, 2009.
23. Mohammad Azadeh. *Fiber Optics Engineering*. New York: Springer Science+Business Media, LLC, 2009.

24. Michael Bass and Eric W. Van Stryland (Eds.). *Fiber Optics Handbook: Fiber, Devices, and Systems for Optical Communications*. New York: The McGraw-Hill Companies, Inc., 2002.
25. Shizhuo Yin, Paul B. Ruffin, and Francis T. S. Yu (Eds.). *Fiber Optic Sensors (2nd Edition)*. Boca Raton: Taylor & Francis Group, LLC, 2008.
26. Branko Glišić and Daniele Inaudi. *Fibre Optic Methods for Structural Health Monitoring*. Chichester: John Wiley & Sons Ltd, 2007.
27. J-P Goure and I Verrier. *Optical Fibre Devices*. Bristol: IOP Publishing Ltd, 2002.
28. Raymond M. Measures. *Structural Monitoring with Fiber Optic Technology*. San Diego: Academic Press, 2001.
29. Ray Tricker. *Optoelectronic and Fiber Optic Technology*. Oxford: Newnes, 2002.

8
CHAPTER

POLARIZATION OF LIGHT

LINEAR POLARIZATION

Let us consider a plane harmonic electromagnetic wave where for its electric **E** and magnetic **H** field the following is valid:

$$\begin{aligned} \mathbf{E}(\mathbf{r},t) &= \mathbf{E}_0 \exp[j(\mathbf{k}\cdot\mathbf{r}-\omega t)] \\ \mathbf{H}(\mathbf{r},t) &= \mathbf{H}_0 \exp[j(\mathbf{k}\cdot\mathbf{r}-\omega t)] \end{aligned} \quad (8.1)$$

In the case when $\mathbf{E}_0$ and $\mathbf{H}_0$ are constant real vectors one can say that we have a linearly polarized light. Figure 8.1. shows the diagram of a linearly polarized light where, as we know from the theory, the electric and magnetic field are mutually perpendicular. It is in tradition in optics to designate the direction of the electric field as the direction of polarization.

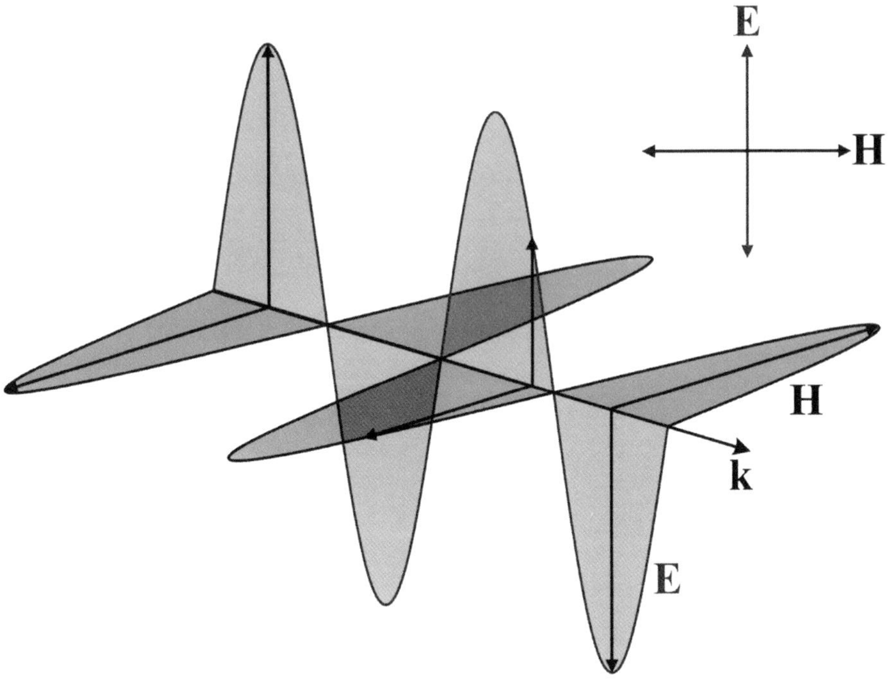

Figure 8.1. Linearly polarized light.

In the quite opposite case, i.e. when we have a natural, or so-called unpolarized light the instantaneous polarization of light fluctuates rapidly in a random manner. In order to obtain a polarized light from, for example an unpolarized light, one can use a linear polarizer. One can find several kinds of linear polarizers. However, the most efficient one are those that are based on the double refraction principle. Another way is to use the phenomenon of anisotropic absorption or so-called dichroism. If a dichoric material is used one component of polarization is more strongly absorbed then the other thus producing a polarized light. One of the mostly common dichroic materials is a natural crystal tourmaline which is not very efficient.

The polarizer transmission axis defines the direction of the electric field vector which is transmitted through the polarizer with little or no loss. A light wave with the electric field vector perpendicular to the polarizer transmission axis is absorbed or attenuated. One can define an ideal polarizer as one that is completely transparent to the light that is polarized in the direction of the transmission axis, and completely opaque to the light that is polarized in the direction perpendicular the transmission axis.

If the unpolarized light impinges the linear polarizer then the incident electric field **E** can be decomposed into two mutually orthogonal components $\mathbf{E}_1$ and $\mathbf{E}_2$ as presented in figure 8.2., where the component $\mathbf{E}_1$ is along the polarizer transmission axis.

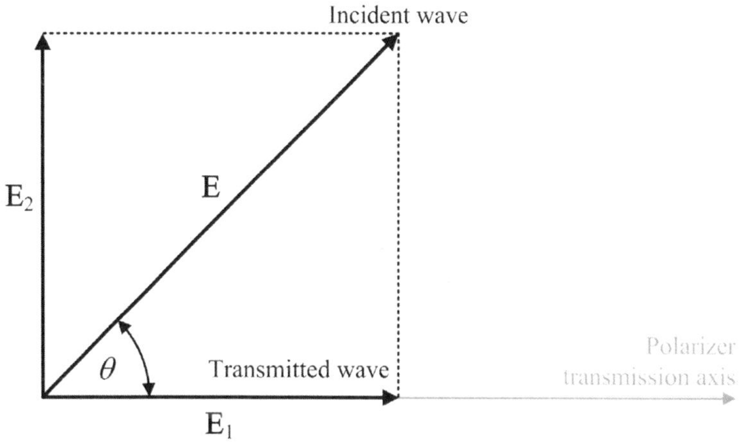

Figure 8.2. Incident and the transmitted electric filed relationship for a linear polarizer.

If the incident electric field makes an angle θ with the transmission axis then the corresponding transmitted electric field magnitude is given by:

$$E_1 = E\cos\theta, \qquad (8.2)$$

where the transmitted irradiance I_1, being proportional to the square of the electric field, is thus given by:

$$I_1 = I\cos^2\theta, \qquad (8.3)$$

where I is the incident beam irradiance. The relation given by equation (8.3) is also known as the Malus's law, which is named after Étienne-Louis Malus. In the case of an unpolarized light, the angle θ can take any value from the range $0 \leq \theta \leq 2\pi$ with equal probability. Therefore, the transmitted irradiance of an unpolarized light is proportional to the incident irradiance by a factor which is equal to the average value of $\cos^2\theta$, i.e. ½.

When the light is partially polarized it can be considered as a mix of polarized and unpolarized light. The degree of polarization one can define as the fraction of the total irradiance which is polarized, i.e. as:

$$P = \frac{I_{\mathrm{P}}}{I_{\mathrm{P}} + I_{\mathrm{UP}}}, \qquad (8.4)$$

where I_{P} and I_{UP} are the irradiances of the polarized and unpolarized light, respectively. One can present the degree of polarization in the following form:

$$P = \frac{I_{\max} - I_{\min}}{I_{\max} + I_{\min}}, \qquad (8.5)$$

where $I_{\max}$ and $I_{\min}$ correspond to the irradiance of the light which is transmitted through a linear polarizer when the polarizer is completely turned around thus covering the whole range of 360°.

CIRCULAR AND ELLIPTIC POLARIZATION

Let us consider the plane harmonic electromagnetic wave consisted of two linearly polarized waves with equal amplitudes, orthogonally polarized to each other. The phase difference between the waves is π/2. If we choose that the electric field vectors are placed along x- and y-axis then we have the following:

$$\begin{aligned} \mathbf{E}_x(z,t) &= \mathbf{i}_x E_0 \cos(k - \omega t) \\ \mathbf{E}_y(z,t) &= \mathbf{i}_y E_0 \sin(k - \omega t), \end{aligned} \qquad (8.6)$$

where E_0 is the field amplitude, $\mathbf{i}_x$ and $\mathbf{i}_y$ unit vectors along x- and y-axis, respectively. The overall electric filed is the vector sum of two components, thus we have:

$$\begin{aligned} \mathbf{E}(z,t) &= \mathbf{E}_x(z,t) + \mathbf{E}_y(z,t) \\ &= E_0\left[\mathbf{i}_x \cos(k - \omega t) + \mathbf{i}_y \sin(k - \omega t)\right]. \end{aligned} \qquad (8.7)$$

The electric field, described with the equation (8.7), can be interpreted as the electric field of a single wave where the electric field vector rotates with the angular frequency ω having at a given point the constant magnitude. This type of wave has a circular polarization, which is presented in figure 8.3. The signs of the corresponding terms in the equation (8.7) define the clockwise rotation of the electric field vector at a given point in space when viewed against the direction of propagation. Such a wave is then right circularly polarized.

If the sign of a second term is changed, the rotation direction is also changed giving a rise to the counterclockwise rotation direction as viewed against the direction of propagation. Such a wave is then left circularly polarized.

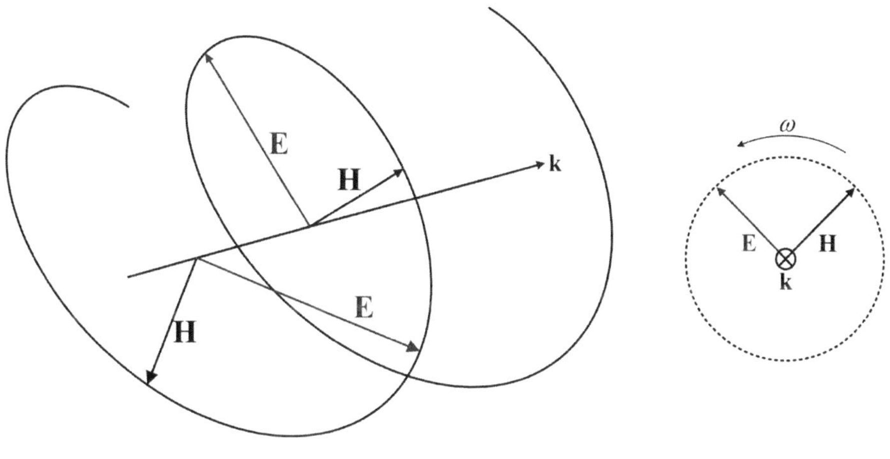

Figure 8.3. Circularly polarized light.

In the complex notation one can define the circularly polarized electrical field as:

$$\mathbf{E}(z,t)=\mathbf{i}_x E_0 \exp[j(kz-\omega t)]+\mathbf{i}_y E_0 \exp\left[j\left(kz-\omega t\pm\frac{\pi}{2}\right)\right]. \qquad (8.8)$$

Bearing in mind the identity: $\exp(j\pi/2)=j$ equation (8.8) becomes:

$$\mathbf{E}(z,t)=E_0(\mathbf{i}_x \pm j\mathbf{i}_y)\exp[j(kz-\omega t)]. \qquad (8.9)$$

One can easily check that the real part of the last expression represents the equation (8.7) where the sign of the imaginary part defines the rotation direction, where the minus sign represents the right circular polarization and the plus sign the left circular polarization.

If the components of the field given by equation (8.7) are not of the same amplitude, i.e.:

$$\mathbf{E}(z,t) = \mathbf{i}_x E_{0x} \cos(kz - \omega t) + \mathbf{i}_y E_{0y} \sin(kz - \omega t), \quad (8.10)$$

where it is fulfilled $E_{0x} \neq E_{0y}$, the resulting vector of electrical filed also rotates at a given point in space where the its magnitude changes in the way that the end of the vector describes an ellipse, as illustrated in figure 8.4. In this case we have an elliptic polarization. If we introduce the so called complex vector amplitude $\mathbf{E}_0$ as follows:

$$\mathbf{E}_0 = \mathbf{i}_x E_{0x} + j\mathbf{i}_y E_{0y}, \quad (8.11)$$

then the corresponding wave function is given by:

$$\mathbf{E}(z,t) = \mathbf{E}_0 \exp[j(kz - \omega t)]. \quad (8.12)$$

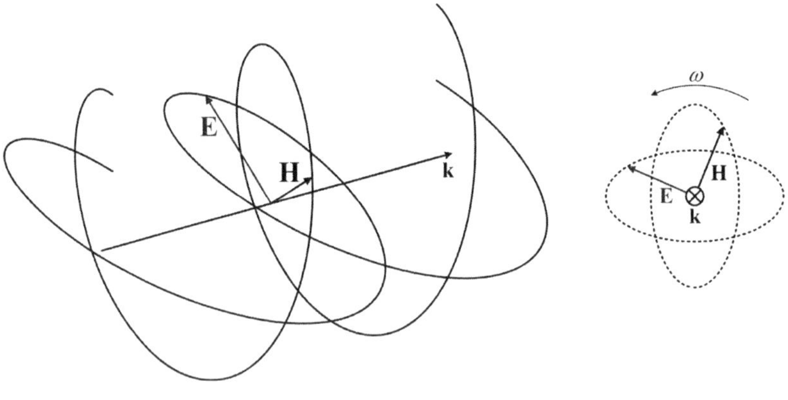

Figure 8.4. Elliptically polarized light.

The expression presented in equation (8.12) represents any type of polarization. If $\mathbf{E}_0$ is real the linear polarization is obtained and if it is complex we have the elliptic polarization. In the special case of elliptic polarization when the real and imaginary parts are equal one obtains the circular polarization.

MATRIX REPRESENTATION OF POLARIZATION

As shown in the previous section, the complex amplitude of the plane harmonic wave can be represented in a more general way as:

$$\mathbf{E}_0 = \mathbf{i}_x E_{0x} + \mathbf{i}_y E_{0y}, \quad (8.13)$$

where E_{0x} and E_{0y} can be both complex. Therefore, they can be represented in the exponential form as:

$$E_{0x} = |E_{0x}|\exp(j\varphi_x)$$
$$E_{0y} = |E_{0y}|\exp(j\varphi_y). \quad (8.14)$$

There is a more convenient notation in the matrix form of the above presented pair of complex amplitudes known as the Jones vector:

$$\begin{bmatrix} E_{0x} \\ E_{0y} \end{bmatrix} = \begin{bmatrix} |E_{0x}|\exp(j\varphi_x) \\ |E_{0y}|\exp(j\varphi_y) \end{bmatrix}. \quad (8.15)$$

The normalized form of the Jones vector can be obtained by dividing by the square root of the sum of the squares of the two modules, i.e. by $\sqrt{|E_{0x}|^2 + |E_{0y}|^2}$.

Calculation with Jones matrix is rather simple when in order to determine the polarization of the resultant wave field of two or more polarized waves one should just add the corresponding Jones vectors of each field. For example, if we have two circularly polarized waves where one have the right circular polarization and the other the left circular polarization, the resultant polarization is obtained according to:

$$\begin{bmatrix} 1 \\ j \end{bmatrix} + \begin{bmatrix} 1 \\ -j \end{bmatrix} = \begin{bmatrix} 2 \\ 0 \end{bmatrix} = 2\begin{bmatrix} 1 \\ 0 \end{bmatrix}, \quad (8.16)$$

i.e. as the linearly polarized wave has been obtained.

Linear polarization of the wave can be obtained by simple transmission of the light through the linear polarizer. However, the circular polarization one can obtain by introducing a phase shift of $\pi/2$ between two orthogonal components of the linearly polarized light. The quarter-wave plate is such a device that allows phase shift between these field components. This plate is made of double refracting glasses such as calcite or mica.

Such glasses have the property that the index of refraction differs for different polarizations of light. One can cut the doubly refracting crystal into slabs where the axis of maximum index of refraction n_s (the slow axis) and the axis of minimum index of refraction n_f (the fast axis) are perpendicular to each other. In order to make a quarter-wave plate from double refracting crystal one must adjust the thickness of the slab the optical path difference between two axes performs a mutual phase shift of exactly π/2. This condition is fulfilled when the thickness of the slab is given by:

$$d = \frac{\lambda}{4(n_s - n_f)}, \quad (8.17)$$

where λ is the wavelength of the light in vacuum. The optical setup for obtaining circularly polarized light from a linearly polarized light is presented in figure 8.5. The quarter-wave plate is placed in a way that the angle between the direction of the linearly polarized light and the fast (slow) axis is set to be 45°. The direction of the rotation of circularly polarized light depends on the rotation position of the quarter-wave plate. If the angle between the linearly polarized light direction and the quarter-wave plate axes is different than ±45° or ±135° one obtain elliptic polarization.

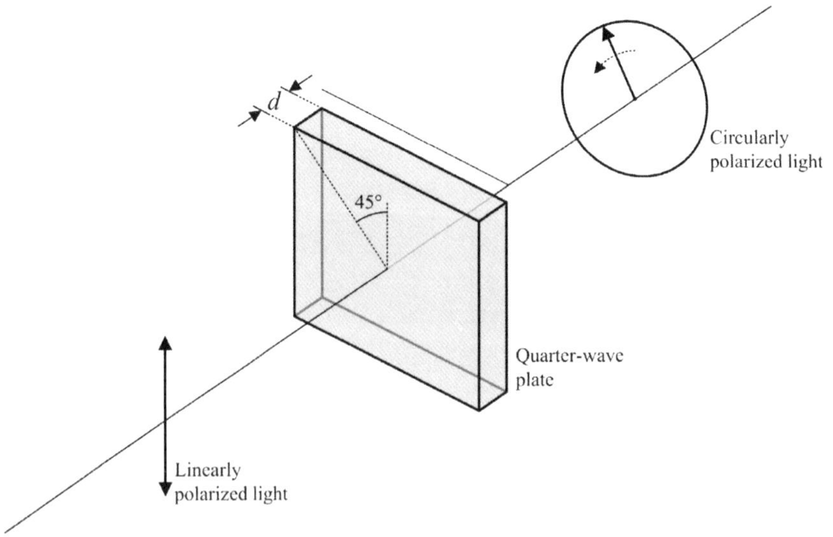

Figure 8.5. Quarter-wave plate.

Matrix notation is also convenient for computing the effect a linear optical element or a chain of such elements, which is inserted in a light beam and alters the polarization of the light. Each optical element can be represented as a 2×2 matrix called Jones matrix. In this way can be represented the linear polarizers, circular polarizers, phase retarders, isotropic phase changers, and isotropic absorbers. The corresponding matrices for such elements are presented in Table 8.1.

Table 8.1. Jones matrices of some linear optical elements

Optical element		Jones matrix
Linear polarizer	Horizontal transmission axis	$\begin{bmatrix} 1 & 0 \\ 0 & 0 \end{bmatrix}$
	Vertical transmission axis	$\begin{bmatrix} 0 & 0 \\ 0 & 1 \end{bmatrix}$
	Transmission axis at ±45°	$\frac{1}{2}\begin{bmatrix} 1 & \pm 1 \\ \pm 1 & 1 \end{bmatrix}$
Quarter-wave plate	Vertical fast axis	$\begin{bmatrix} 1 & 0 \\ 0 & -j \end{bmatrix}$
	Horizontal fast axis	$\begin{bmatrix} 1 & 0 \\ 0 & j \end{bmatrix}$
	Fast axis at ±45°	$\frac{\sqrt{2}}{2}\begin{bmatrix} 1 & \pm j \\ \pm j & 1 \end{bmatrix}$
Half-wave plate	Vertical or horizontal fast axis	$\begin{bmatrix} 1 & 0 \\ 0 & -1 \end{bmatrix}$
Isotropic retarder		$\begin{bmatrix} \exp(j\phi) & 0 \\ 0 & \exp(j\phi) \end{bmatrix}$

Relative phase changer		$\begin{bmatrix} \exp(j\phi_x) & 0 \\ 0 & \exp(j\phi_y) \end{bmatrix}$
Circular polarizer	Right	$\frac{1}{2}\begin{bmatrix} 1 & j \\ -j & 1 \end{bmatrix}$
	Left	$\frac{1}{2}\begin{bmatrix} 1 & -j \\ j & 1 \end{bmatrix}$

The corresponding Jones matrices can be used in the following way. If the vector of the incident light is $\begin{bmatrix} A \\ B \end{bmatrix}$ and the vector of the exiting light is $\begin{bmatrix} A' \\ B' \end{bmatrix}$ then we have the following:

$$\begin{bmatrix} A' \\ B' \end{bmatrix} = \begin{bmatrix} a & b \\ c & d \end{bmatrix} \begin{bmatrix} A \\ B \end{bmatrix} = \mathbf{J} \begin{bmatrix} A \\ B \end{bmatrix}, \tag{8.18}$$

where **J** is the Jones matrix of the optical element. In the case when we have a chain of n optical elements then the Jones vector of the light emerging from all the elements is given by:

$$\begin{bmatrix} A' \\ B' \end{bmatrix} = \begin{bmatrix} a_n & b_n \\ c_n & d_n \end{bmatrix} \cdots \begin{bmatrix} a_2 & b_2 \\ c_2 & d_2 \end{bmatrix} \begin{bmatrix} a_1 & b_1 \\ c_1 & d_1 \end{bmatrix} \begin{bmatrix} A \\ B \end{bmatrix}. \tag{8.19}$$

For two waves, where the complex vector amplitudes $\mathbf{E}_1$ and $\mathbf{E}_2$ represent their states of polarizations, is said to be orthogonally polarized if it is fulfilled:

$$\mathbf{E}_1 \cdot \mathbf{E}_2^* = 0. \tag{8.20}$$

In the case of a linearly polarized light their orthogonality means that the fields are polarized at right angle to one other. For a circularly polarize light right circular and left circular polarizations are mutually orthogonal. In general case, the waves are orthogonal if it is fulfilled:

$$A_1 \cdot A_2^* + B_1 \cdot B_2^* = 0. \tag{8.21}$$

where $\begin{bmatrix} A_1 \\ B_1 \end{bmatrix}$ and $\begin{bmatrix} A_2 \\ B_2 \end{bmatrix}$ represent the corresponding Jones vectors of the waves.

REFLECTION AND REFRACTION

In order to investigate the reflection and refraction phenomena at a plane boundary between two dielectric surfaces we will observe the situation as presented in figure 8.6. With **E** we will denote the amplitude of the electric field of a plane harmonic wave, which impinges the plane boundary that separates two dielectric media with the corresponding indices of refraction n_1 and n_2, and with **E'** and **E''** we will denote the amplitudes of the reflected and transmitted waves, respectively. According to the Maxwell's equation one has:

$$\mathbf{H} = \frac{1}{\mu\omega}\mathbf{k} \times \mathbf{E}$$

$$\mathbf{H}' = \frac{1}{\mu\omega}\mathbf{k}' \times \mathbf{E}'$$

$$\mathbf{H}'' = \frac{1}{\mu\omega}\mathbf{k}'' \times \mathbf{E}'' \quad , \qquad (8.22)$$

where **H**, **H'**, and **H''** are the corresponding magnetic fields. Further, we must differ two cases. The first case corresponds to the electrical field parallel to the boundary plane, i.e. perpendicular to the plane of the drawings presented in figure 8.6.(a). This case is called the transverse electric or TE polarization. The second case corresponds to the magnetic field parallel to the boundary plane, i.e. perpendicular to the plane of the drawings presented in figure 8.6.(b). This case is called the transverse magnetic or TM polarization.

According to the well-known boundary conditions, where the tangential electric and magnetic fields must be the same at both sides of the boundary, we have **E''** = **E**+**E'** for the TE polarization and **H''** = **H**-**H'** for TM polarization, so one has:

$$E + E' = E''$$
$$-H\cos\theta + H'\cos\theta = -H''\cos\phi$$
$$-kE\cos\theta + k'E'\cos\theta = -k''E''\cos\phi, \tag{8.23}$$

for the TE polarization and:

$$H - H' = H''$$
$$kE - k'E' = k''E''$$
$$E\cos\theta + E'\cos\theta = E''\cos\phi, \tag{8.24}$$

for the TM polarization.

The corresponding coefficients of reflection r_s and r_p and the coefficients of transmission t_s and t_p are defined as the following ratios:

$$r_s = \left.\frac{E'}{E}\right|_{TE} \qquad r_p = \left.\frac{E'}{E}\right|_{TM}$$
$$t_s = \left.\frac{E''}{E}\right|_{TE} \qquad t_p = \left.\frac{E''}{E}\right|_{TM}. \tag{8.25}$$

By taking into consideration the corresponding indices of refraction we obtain:

$$r_s = \frac{n_1\cos\theta - n_2\cos\phi}{n_1\cos\theta + n_2\cos\phi}$$
$$r_p = \frac{-n_2\cos\theta - n_1\cos\phi}{n_2\cos\theta + n_1\cos\phi}. \tag{8.26}$$

According to the Snell's law we have $n_1\sin\theta = n_2\sin\phi$, so the equation (8.25) can be rewritten in the form:

$$r_s = -\frac{\sin(\theta-\phi)}{\sin(\theta+\phi)}, \quad t_s = \frac{2\cos\theta\sin\phi}{\sin(\theta+\phi)}$$
$$r_p = -\frac{\tan(\theta-\phi)}{\tan(\theta+\phi)}, \quad t_p = \frac{2\cos\theta\sin\phi}{\sin(\theta+\phi)\cos(\theta-\phi)}, \tag{8.27}$$

where the above presented equations are known at the Fresnel's equations.

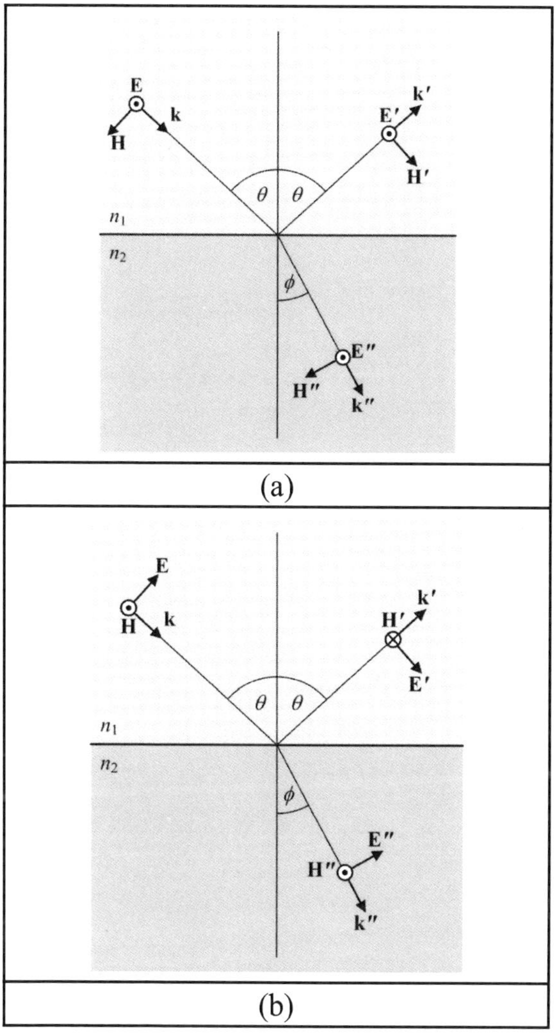

Figure 8.6. Wave vectors and the corresponding fields for (a) TE and (b) TM polarization.

The corresponding reflectances, which are defined as the energetic ratios of the reflected and incident beams, are proportional to the absolute square of the field amplitudes ratio:

$$R_s = |r_s|^2, \quad R_p = |r_p|^2 \quad (8.28)$$

In figure 8.7 are presented the variations of the reflection coefficients r_s, r_p, R_s, and R_p with the angle of incidence θ which were calculated from equation (8.25) where the Snell's law was taken into account in the following way:

$$r_s = \frac{\cos\theta - \sqrt{n^2 - \sin^2\theta}}{\cos\theta + \sqrt{n^2 - \sin^2\theta}}$$

$$r_p = \frac{-n^2\cos\theta + \sqrt{n^2 - \sin^2\theta}}{n^2\cos\theta + \sqrt{n^2 - \sin^2\theta}}, \qquad (8.29)$$

where $n = n_2/n_1$ is valid.

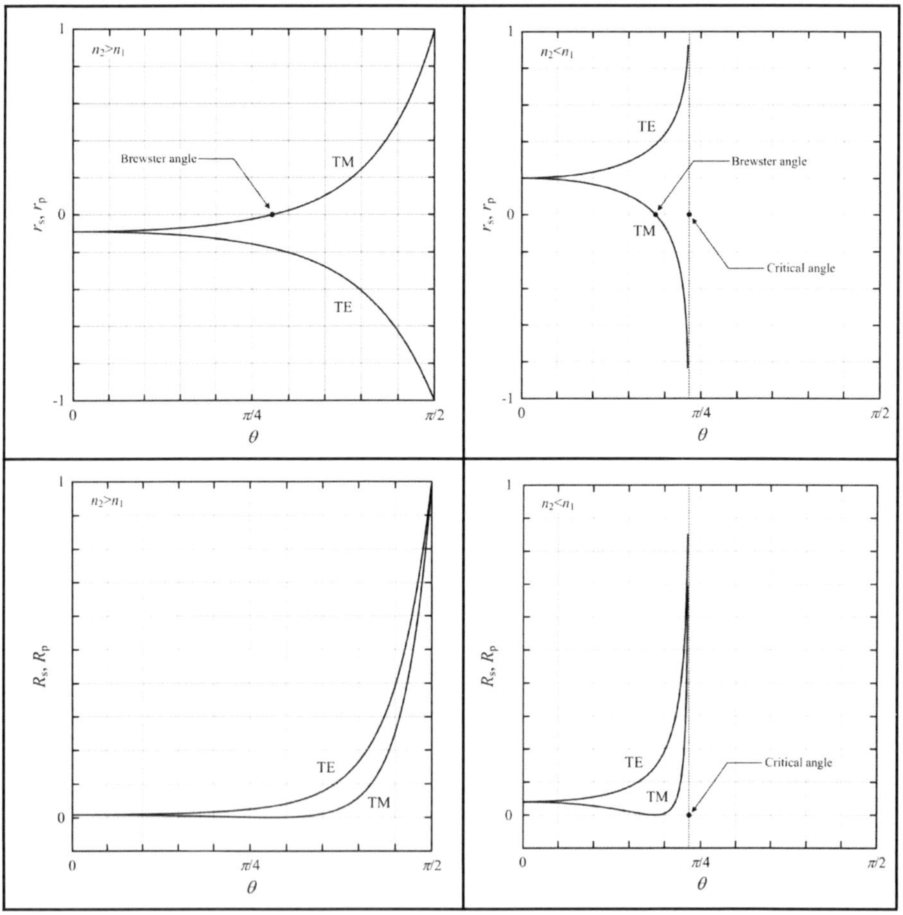

Figure 8.7. Reflection coefficients the variations with the angle of incidence.

THE BREWSTER ANGLE

If we return back to the last equation one can notice that for a certain angle of incidence for TM polarization there is no back reflected light. This particular angle of incidence depends on the medium indices of refraction as:

$$\theta_B = \arctan n \qquad (8.30)$$

This angle is called the polarizing angle or the Brewster angle. For a glass with the index of refraction 1.5 we have for the case of external reflection from the air to the glass $\theta_B = \arctan 1.5 \approx 5\ ^\circ$. In the case of the internal reflection from the glass to the air we have $\theta_B = \arctan(1/1.5) \approx 3\ ^\circ$. So, when unpolarized light impinges the surface at the Brewster angle the reflected light is linearly polarized with the electric field vector transverse to the plane of incidence. However, only a small fraction of light (1.5% for the index of refraction of 1.5) is reflected and thus the polarization is not very efficient.

If a linearly polarized beam of light, in the TM mode, impinges at Brewster's angle on a glass slab, as presented in figure 8.8., there will be no reflection from the first glass-air interface, as well as from the second air-glass interface. The result is that the light is totally transmitted through the window thus eliminating any back reflection. This perfect window is called the Brewster's window, which is extensively used in the laser applications.

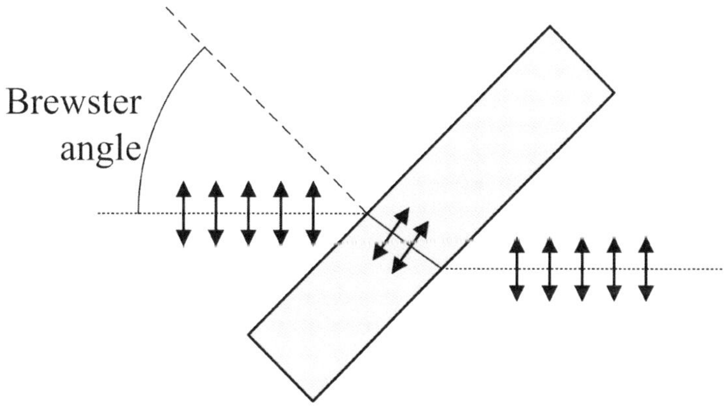

Figure 8.8. Brewster window.

REFERENCES

1. Rosario Martínez-Herrero, Pedro M. Mejías, and Gemma Piquero. *Characterization of Partially Polarized Light Fields*. Berlin: Springer-Verlag Berlin Heidelberg, 2009.
2. David Pye. *Polarised Light in Science and Nature*. Bristol: IOP Publishing Ltd, 2001.
3. Andrea Galtarossa and Curtis R. Menyuk (Eds.). *Polarization Mode Dispersion*. New York: Springer Science+Business Media, Inc., 2005.
4. Jay N. Damask. *Polarization Optics in Telecommunications*. New York: Springer Science+Business Media, Inc., 2005.
5. Toralf Scharf. *Polarized Light in Liquid Crystals and Polymers*. Hoboken: John Wiley & Sons, Inc., 2007.
6. Dennis Goldstein. *Polarized Light (2nd Edition, Revised and Expanded)*. New York: Marcel Dekker, Inc., 2003.
7. Edward Collett. *Polarized Light: Fundamentals and Applications*. New York: Marcel Dekker Inc., 1992.
8. Prasad L. Polavarapu (Ed.). *Principles and Applications of Polarization-division Interferometry*. Chichester: John Wiley & Sons, Ltd, 1998.

9 CHAPTER

OPTICAL DETECTION

The previous chapters describe the optical phenomena whether they are based on the geometrical or wave optics. Roughly speaking we can place an arbitrary optical system into several categories such as imaging optics, non-imaging optics, fiber-optics, etc. Regardless the purpose of the particular optical system, at the end there is an optical detector. Typically, optical detector converts an optical signal into more convenient form for signal extraction and further processing. As an example of an imaging optical system can serve the camera where at the end one has a two-dimensional array of optical detector that converts the image into electrical signal which is more convenient for further processing. In the case of a non-imaging optics we can refer to the collectors of the solar radiation where at the end there is a photovoltaic solar cell that converts the captured optical energy into electrical energy that is, once again, more convenient for further distribution. Fiber-optic communication systems and optical sensing systems always at their end have an optical detector where the optical data stream is converted into an electrical data stream or an electrical signal proportional to the measured physical quantity, respectively. However, the captured light isn't always transformed in its electrical counterpart, yet again the most detectors employ this type of optical detection. Table 9.1. lists the most common detection techniques that are associated with the detection of optical signals.

Table 9.1. Photodetection techniques

Thermal Effects	Wave Interaction Effects	Photon (Quantum) Effects
Thermoelectric effect	Parametric down-conversion	Photoconductors
Pyromagnetic effect	Parametric up-conversion	Photoemissives
Pyroelectric effect	Parametric amplifiers	Photovoltaics
Liquid crystals		
Bolometers		

Optical detectors based on the thermal effect employ the detector temperature change when absorb the optical radiation. With the detector temperature change alters some other device parameter, usually its resistance that can be externally sensed. Typical example of the thermoelectric detectors is the bolometer where the bolometer resistance changes with the temperature change. Also we have the pyroelectric detectors where the detector capacitive charge alters with the temperature changes.

Optical detectors that are based on the wave interaction effect employ the interaction of the lightwave and a nonlinear material in order to form the sum or difference frequency of the lightwaves. The wave interaction is the basis for the construction of optical parametric amplifiers and frequency doubles. However, efficient optical parametric amplifiers are relatively complex to implement. Therefore, they haven't been widely used in the optical communication systems.

The third group of optical detectors that are based on the photon (quantum) effects is the most popular one. These types of detectors employ the absorption of the photons in order to convert them into the corresponding photocarrier stream. The generation of a photocarrier corresponds to the electron-hole pair formation within the photodetector material. The photocarriers transport the charge thus forming an electrical photocurrent which can be further processed by using conventional electronics.

PHOTON COUNTING

Optical signal that impinges an optical detector can be modeled as a stream of incident photons. In the case of a monochromatic lightwave each photon carries the same amount of energy which is independent on the of incident beam. However, the photon rate varies with the power of the optical signal. High energetic content of an optical signal corresponds to the high photon arrival rate. Therefore, from the receiver point of view the most important parameter is the number of the received photons during particular time interval. Essentially, the detection process is an events counting process known as the photon counting process as illustrated in figure 9.1.

A constant power, monochromatic beam, which consists of a photon stream with photon each carrying energy $E = h\nu$, illuminates an ideal photon to photon carrier converter, as it is presented in figure 9.1.(a). The photons arrive at discrete points in time as presented in figure 9.1.(b). Such formed point process is random.

An ideal photon to photocarrier converter is a device with infinite bandwidth which converts every incident photon without an exception into a photoelectron. Further, an ideal electric counter counts the generated photocarriers. When the counter reset occurs after specified time interval the number of photons has been counted by the detector within each time interval, as illustrated in figure 9.1.(c). If the light source power is modulated the number of counted received photons within the defined time interval also varies in the same manner.

It is well known that the photon generation of an optical source (transmitter) is a random process. Therefore, the point process at the receiver site is as well random process. The number of photocarriers generated within an ideal photodetector illuminated by a constant power monochromatic light is described by the Poisson distribution as:

$$P(n|t) = \frac{(rt)^n \exp(-rt)}{n!} \quad . \tag{9.1}$$

where $P(n|t)$ is the probability of detecting n photons within the time interval t and r is the mean photon arrival rate.

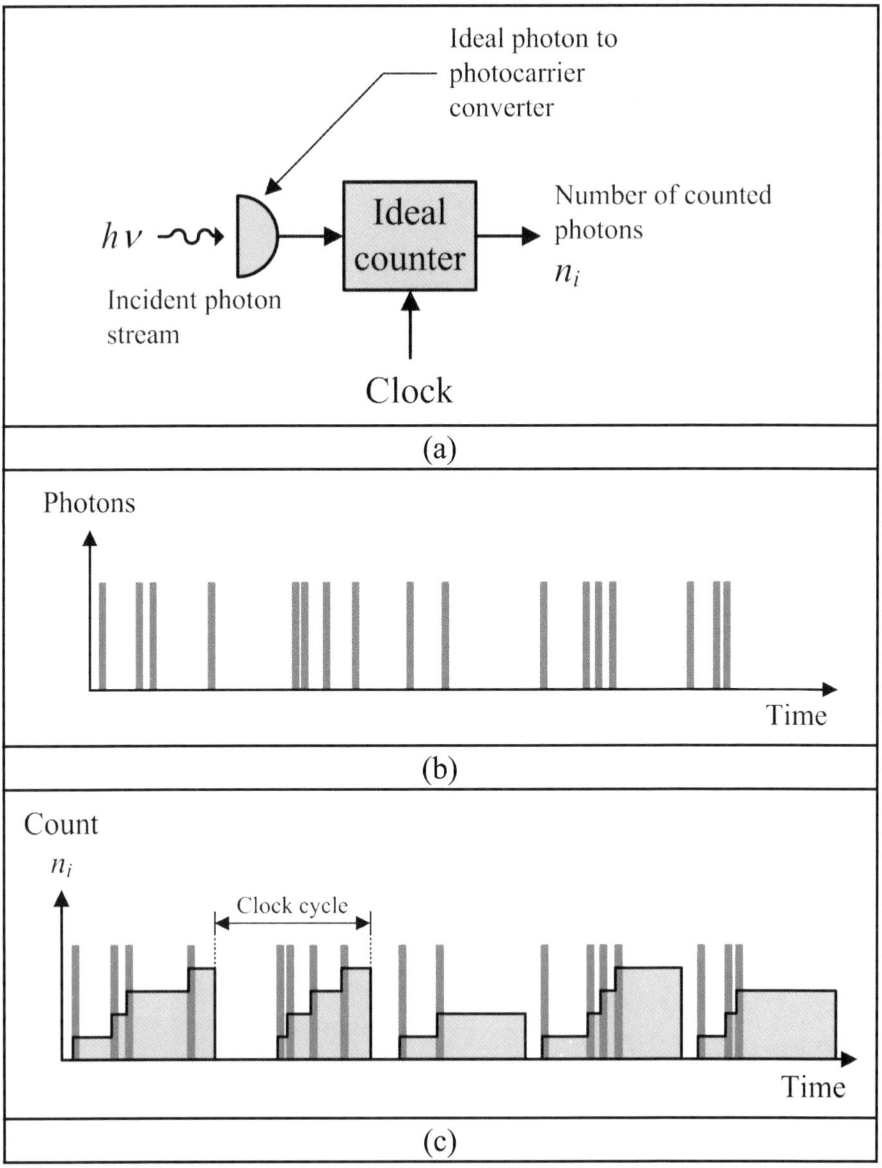

Figure 9.1. Photon counting: (a) ideal photon detector and counter, (b) photon arrivals, and (c) photon count.

For the fixed observation time *t* and the mean photon arrival rate *r* we have:

$$P(n) = \frac{s^n \exp(-s)}{n!}, \qquad (9.2)$$

where $s = t$ is valid. The examples of the Poisson distributions are presented in figure 9.2. for the mean arrival rates of s = 5, s = 20, and s = 80 photons per observation time interval.

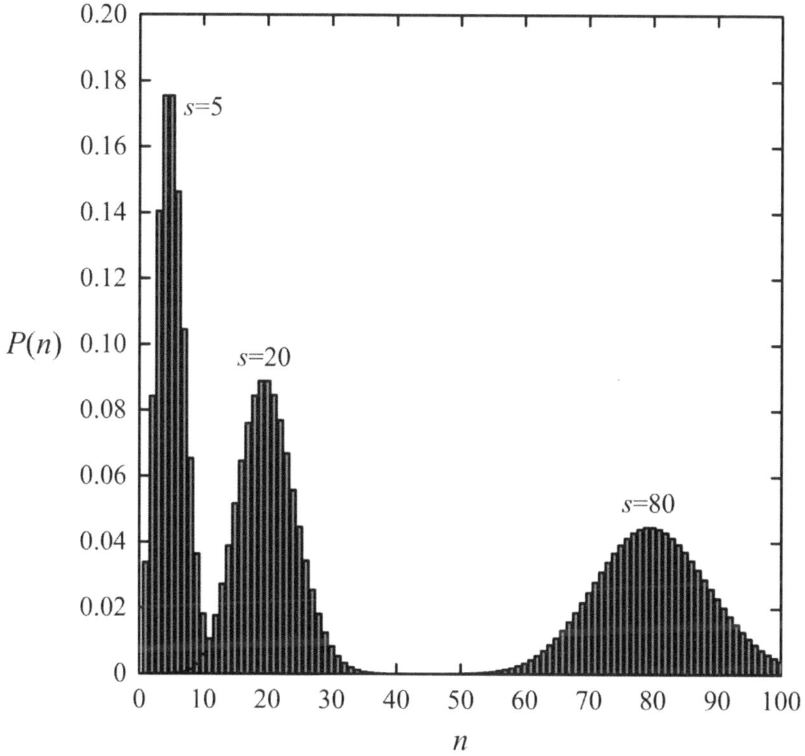

Figure 9.2. Poisson distribution examples.

Poisson distribution represents an accurate description of many processes with the common features that they appear as a series of random independent events and where the probability that the event will occur during certain time interval is proportional to the time interval length. This distribution as well describes the process of the photocarrier generation within the photodetector when illuminated with coherent, monochromatic light such as is generated by laser. However, the photons of the light, generated by the incoherent sources such as incandescent and fluorescent light bulbs, the Sun, and the amplified spontaneous emission from the optical amplifiers don't obey the Poisson distribution. The distribution which describes such random process, where the photodetector is illuminated by the incoherent narrowband light is known as the Bose-Einstein distribution which is given by:

$$P(n) = \left(\frac{1}{1+b}\right)\left(\frac{b}{1+b}\right)^n, \qquad (9.3)$$

where b is again the mean number of incoherent background photons which is defined as:

$$b = \frac{N}{h\nu}, \qquad (9.4)$$

where N is the spectral density of incoherent light specified as:

$$N = \frac{p}{B}, \qquad (9.5)$$

where p is the power of the incoherent source and B is the optical bandwidth of the incoherent source. If the t is the photon counting time interval of the incoherent source the narrowband condition which must be met is $t \le 1/B$.

In the case when the photodetector is illuminated by both sources a coherent signal and a narrowband background component the photon counting statistics obeys the Laguerre distribution given by:

$$P(n) = \left(\frac{1}{1+b}\right)\left(\frac{b}{1+b}\right)^n \exp\left(\frac{s}{1+b}\right) L_n\left[-\frac{s}{b(1+b)}\right], \qquad (9.6)$$

where the mean photon rate of the coherent signal s can be presented as:

$$s = \frac{p_S}{h\nu_S}, \qquad (9.7)$$

where p_S is the optical power of the coherent signal of the frequency ν_S, and $L_n(x)$ is the Laguerre polynomial.

PHOTODETECTION MODELING

The ideal photon counter, as presented in figure 9.1. is impossible to realize in practice. A more realistic model of a real photodetector is presented in figure 9.3. Instead of an ideal photon to photocarrier converter here we have the converter with conversion efficiency η which is lower than unity and an electrical low-pass filter to include into account the finite time response of a real photodetector.

Consequently figure 9.4. represents the corresponding current outputs from the real photon to photocarrier converter. The generated photocurrent is a series of current pulses each carrying the charge of a single electron.

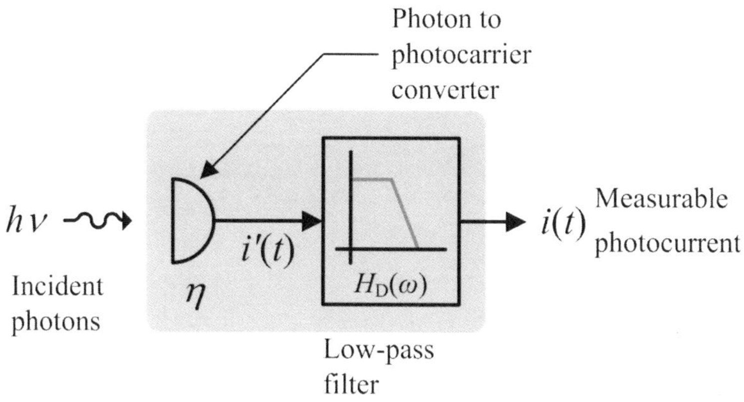

Figure 9.3. Photodetector model.

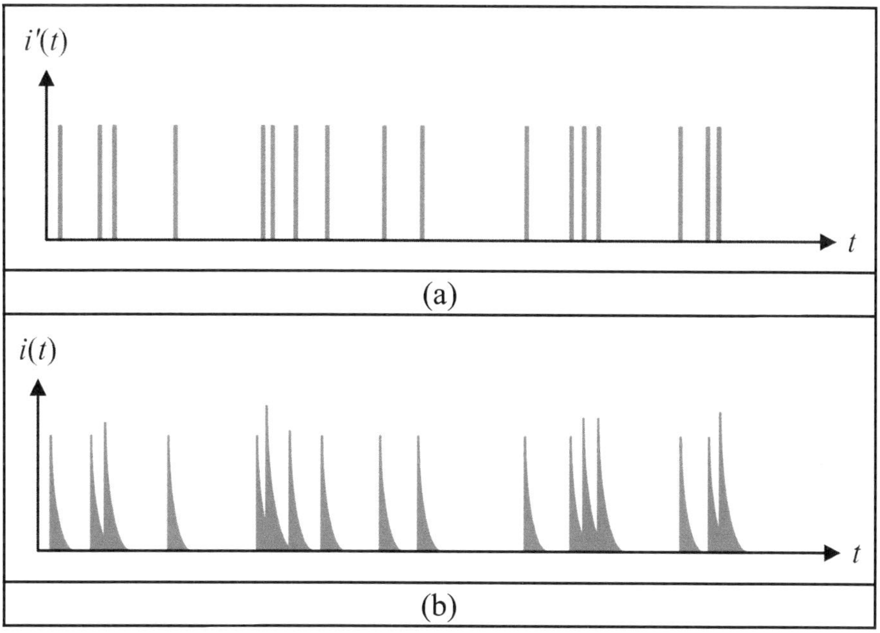

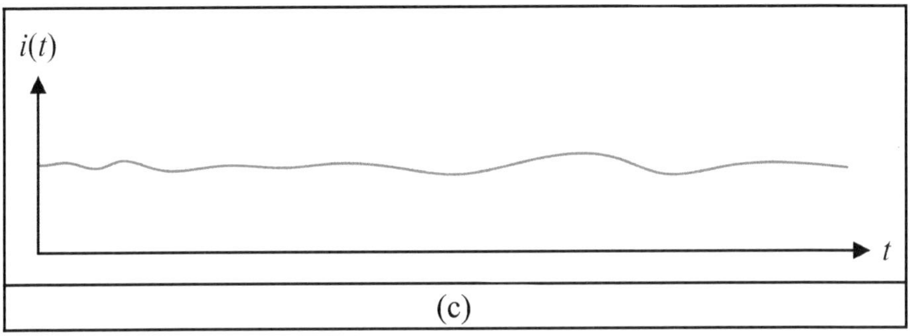

Figure 9.4. Photodetector current characteristics.

How the inherent low-pass filter of the photodetector acts on the current pulses is presented in figure 9.4.(b). The obtained current pulses have a finite time duration which depends on the filter bandwidth. Each pulse still carries the charge equal to the charge of a single electron. When the number of photons per observation time interval increases a continuous photocurrent is obtained as presented in figure 9.4.(c). Baring in mind that the photodetector represents a linear system, in the general case, we can model the photodetector current as a sum of individual photodetector impulse responses as:

$$i(t)=\sum_{k=1}^{n} h_{\mathrm{D}}(t-\tau_k), \qquad (9.8)$$

where $h_{\mathrm{D}}(t)$ represents the photodetector impulse response, n is the total number of generated photocarriers, and τ_k is the random time at which the k-th photocarrier has been generated. Each current pulse carries a charge of a single electron thus we have:

$$\int h_{\mathrm{D}}(t)\mathrm{d}t = q, \qquad (9.9)$$

where q is the electron charge.

The number of photocarriers generated during a particular time interval depends on the optical power captured by the photodetector. By including the variation of the captured optical power with time, the photon rate also changes with the time as:

$$r(t)=\frac{\eta}{h\nu}p(t), \qquad (9.10)$$

where $p(t)$ is the captured optical power and η is the so-called quantum efficiency defined as the ration of the produced photocarriers number and the number of incident photons for which is valid $0 \leq \eta \leq 1$. Baring in mind that the photocurrent is proportional to the photon rate as $i(t) = q\,(t)$ and taking into consideration the equation (9.10) one obtains:

$$i(t) = \frac{\eta q}{h\nu} p(t) = \Re p(t) \qquad (9.11)$$

The term given by:

$$\Re = \frac{\eta q}{h\nu} p(t), \qquad (9.12)$$

is known as the photodetector responsivity. As presented in figure 9.5. the responsivity varies with the wavelength. Responsivity increases with wavelength because there are more photons per unit power at long wavelengths then there are at short wavelengths. Since the amount of photocurrents depends on the photon number not the photon energy, longer wavelengths generate more photocurrent per unit power than do the short wavelengths.

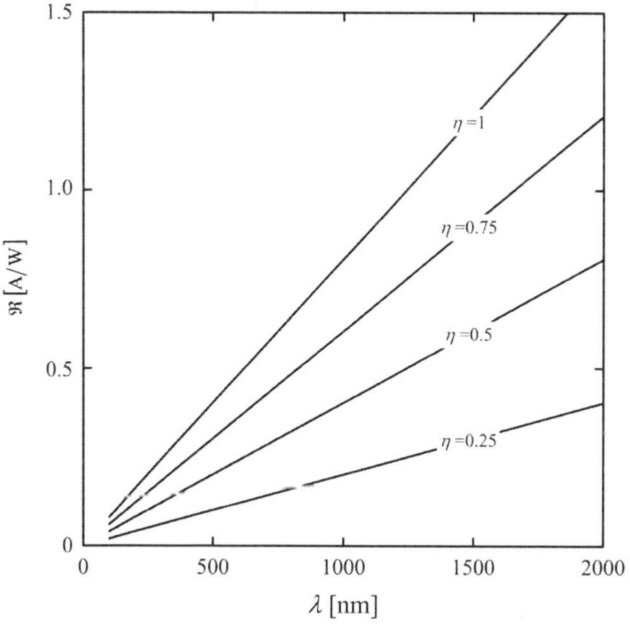

Figure 9.5. Wavelength dependence of the responsivity.

PHOTODETECTORS

The photon effect based photodetector directly generates the photocurrent from the interactions between the photons and the atoms within the detector material. After the penetration into a photodetector the probability that the particular atom will absorb a photon and thus generate free carrier is rather small. However, the number of atoms within the detector is huge so the probability that the photon will be absorbed by some atom is quite high. Photon-effect photodetectors can be designed to fulfill two major requirements such as high sensitivity and wide bandwidth (fast response time). Such photodetectors are typically grouped into four major categories:

- Photomultipliers,
- Photoconductors,
- Photodiodes, and
- Avalanche photodiodes.

Each of these four photon-effect based photodetectors will be described in the following sections.

Photomultipliers

A photomultiplier (PMT) is a vacuum tube that employs the photoelectric effect as well as the secondary emission of electrons in order to provide high current gain. A schematic representation of the photomultiplier tube is given in figure 9.6. When photon reaches the photocathode, which is made of a low work-function metal or semiconductor, then if the photon energy exceeds the work-function the electron will be ejected with a certain probability. After the ejection, electron is accelerated by the electric field formed between the series of electrodes called dynodes. The dynodes are coated with a material that is prone to secondary emission of electrons and are connected in series between the resistors. The resistors R that are connected in series between the dynodes form the gradient of voltage from the high voltage source $+V_{HV}$ to the ground. The dynodes generate additional electrons by the secondary emission. Finally, all the electrons that are generated by the dynodes are captured by the anode. The current multiplication process continues for each dynode in the photomultiplier tube thus providing the current gain that can exceed 10^6.

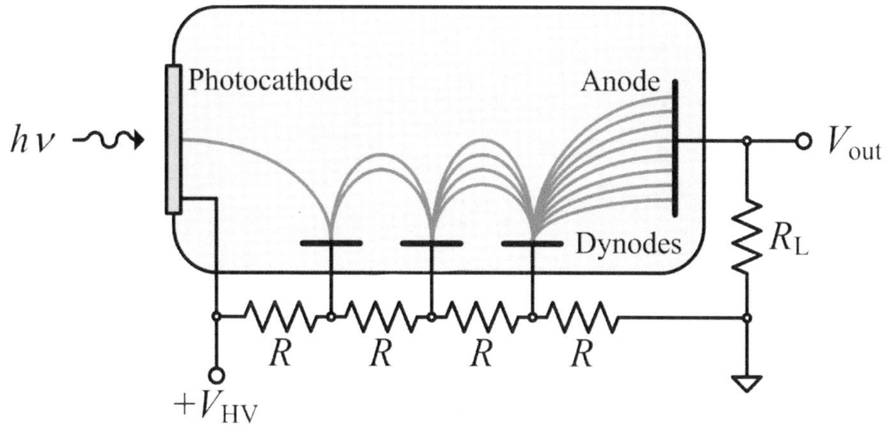

Figure 9.6. Photomultiplier tube.

The probability of detecting incident photon is equal to the quantum efficiency of the photomultiplier tube. Typical photocathode can reach a quantum efficiency of 30% meaning that in average approximately every third photon is detected.

Photomultipliers are usually used for the laboratory instrumentation. Due to their very high current gain the photomultiplier tubes are used to detect extremely low optical signals. PMTs are physically large, require high voltage power supplies and are mechanically fragile.

Photoconductors

Photoconductors employ the photon absorption in semiconductor material. There are three principal absorption processes that can be found in semiconductors, as presented in figure 9.7. The intrinsic band-to-band absorption appears in the cases when the bandgap energy E_g of the material is lower than the photon energy. Therefore, the photon is able to excite the electron from the valence to the conduction band. Such obtained hole and electron constitute a charge carrier. If an external electric field is applied the hole and electron propagate throughout the semiconductor material to the external circuit thus causing photocurrent flow.

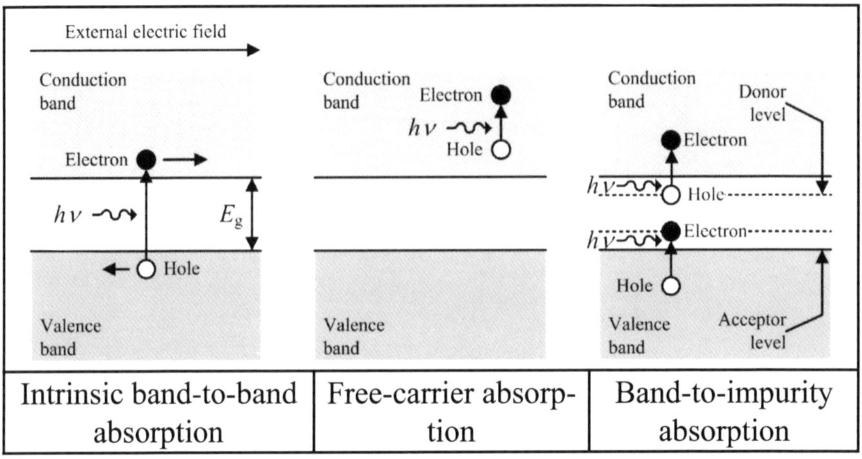

Figure 9.7. Photon absorption mechanisms in semiconductors.

Intrinsic band-to-band absorption represents a dominant absorption mechanism in most semiconductors which are used for the photodetection purposes. In order to trigger the intrinsic band-to-band absorption the following condition must be fulfilled:

$$h\nu > E_g$$

$$\frac{h}{\lambda} > E_g, \qquad (9.13)$$

where $h\nu$ is the photon energy. According to equation (9.13) there is a maximum wavelength where the intrinsic band-to-band absorption still can occur. This maximum wavelength is given by:

$$\lambda_{max} = \frac{h}{E_g}. \qquad (9.14)$$

The bandgap energies and the corresponding maximum wavelengths for typical semiconductor materials used in the photodetection are presented in table 9.2.

Table 9.2. Typical semiconductor materials characteristics

Material	Bandgap E_g [eV]	Maximum wavelength λ_{max} [nm]	Operating spectral range [nm]
Si	1.12	1110	500 – 900
Ge	0.67	1850	900 – 1300
GaAs	1.43	870	750 – 850
$In_xGa_{1-x}As_yP_{1-y}$	0.38 – 2.25	550 – 3260	1000 – 1600

Free-carrier absorption, a secondary effect absorption, occurs when a material absorbs a photon, and a carrier is excited from an already-excited state to another, unoccupied state in the same band. The third absorption mechanism, band-to-impurity absorption is also a secondary effect absorption at near infrared band. Typically, it is used in the mid-infrared region in order to develop photodetectros that are sensitive up to the 30 μm of the radiation wavelength.

A photoconductor detector can have a very simple design based on the semiconductor slab of the bulk semiconducting material with the ohmic (metal) contacts at the ends, as presented in figure 9.8.(a) or it may have an interdigitated structure of the ohmic contacts on the top as presented in figure 9.8.(b). In order to bias the photoconductor a simple DC circuit can be implemented, which generates the corresponding photocurrent I_p proportional to the photoconductor conductance and thus the captured optical power.

The photocurrent of an illuminated photoconductor is influenced by the carrier lifetime and the time required to make their way through the device:

$$I_p = \frac{\eta q}{h\nu}\frac{\tau_C}{\tau_T}p + I_d \quad , \qquad (9.15)$$

where τ_C is the mean carrier lifetime, τ_T transit time between the electrical contacts, p is the captured optical power, and I_d is the dark current.

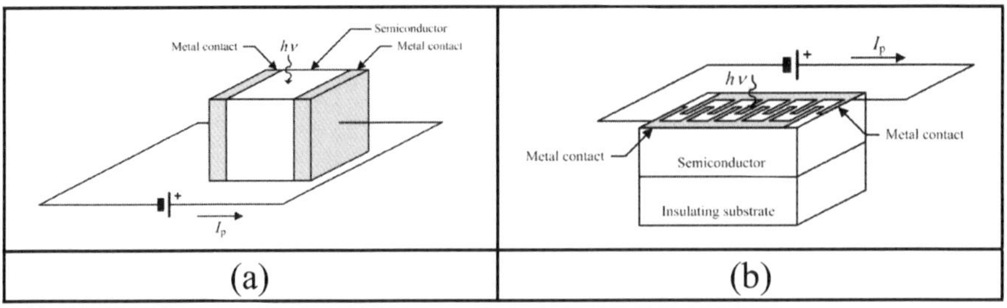

Figure 9.8. Photoconductor design: (a) bulk semiconductor and (b) semiconductor film with an interdigitated electrode structure.

Photodiodes

A semiconductor p-n junction (diode) can be used as a photodetector which is frequently called a photodiode. Similarly like in the case of the photoconductors photodiode relies on the photon absorption in a semiconductor material. Such generated photocarriers are separated by an electric field thus resulting in a photocurrent proportional to the captured optical power. The p-n junction is illustrated in figure 9.9.(a) together with a biasing circuitry. The junction doping profiles are shown in figure 9.9.(b) where the p-type side of the photodiode contains a surplus of holes and which has been doped with the acceptor atoms of a concentration of N_a. The other side of a photodiode is the n-type semiconductor containing a surplus of electrons. The concentration of the donor atoms at the n-side is N_d.

When the junction has been formed, electrons from the n-type material and holes from the p-type material diffuse across the junction in order to equalize the free-carrier density throughout the semiconductor material. Electrons and holes continue to flow until the charge buildup, as shown in figure 9.9.(c), is sufficient to prevent any additional carriers from crossing the junction.

The region in the junction vicinity where the charge buildup occurs is known as the space-charge region or the depletion region. The corresponding energy diagram is shown in figure 9.9.(d) where the energy band diagram illustrates the relationship between the top of the valence band and the bottom of the conductance band across the photodiode. The Fermi level, which defines the energy level with a 50% probability of occupancy, is also shown. Finally, in figure 9.9.(e) is shown the electric field profile in the depletion region.

When a photon is absorbed within the depletion region an electron-hole pair will be formed. The electron and hole will be transported throughout the depletion region by the inherent electric field to the edges of the depletion region. After leaving the depletion region the carriers travel to the terminals of the photodiode thus forming a photocurrent that further flows into the external circuitry.

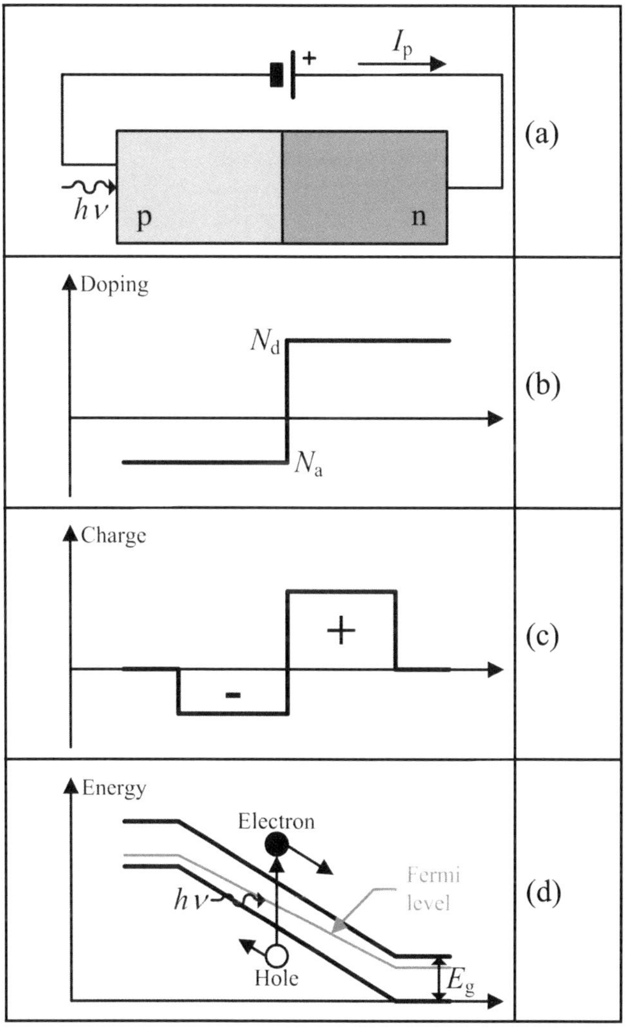

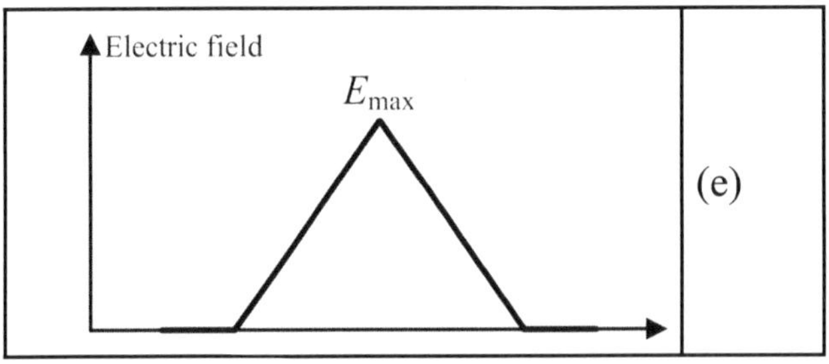

Figure 9.9. Reverse biased p-n diode as a photodetector.

The built-in electric filed is usually supplemented by an external field which biases the photodiode reversely. The stronger the external electric field the wider the depletion region and the more rapidly the free carriers will move toward the region edges. The average time required for a photocarrier to travel across the depletion region is called the transit time and is given by:

$$\tau_T = \frac{l_D}{v_S}, \quad (9.16)$$

where l_D is the depletion region length and v_S is the average carrier saturation velocity. The shorter the transit time of the photodiode the faster is its response to the abrupt illumination.

As it is well known the photodiode is still a diode. The diodes can operate in three modes, unbiased, forward biased, and reverse biased. Figure 9.10. illustrates these three operating modes with three corresponding bias schematics in which the photodiode can operate when connected to the external circuit. These three modes are usually referred to as open circuit or the photovoltaic mode, short circuit or photoconductive mode, and the reverse biased photoconductive mode.

Regardless the operation mode the photodiode current vs. voltage characteristics given by:

$$i_D = I_S\left[\exp\left(\frac{v_D}{V_T}\right)-1\right]+i_L-i_P, \quad (9.17)$$

where I_S is the reverse saturation current, i_L is the photodiode leakage current, v_D is the voltage applied across the photodiode, V_T is the thermal voltage given by:

$$V_T = \frac{k_B T}{q}, \qquad (9.18)$$

where k_B is the Boltzmann constant, T is the photodiode absolute temperature and q is elementary positive charge. The photocurrent i_P from equation (9.17) is given as:

$$i_P = \frac{\eta q}{h\nu} p. \qquad (9.19)$$

The photodiode dark current, which flows through the photodiode when it isn't illuminated, is given by:

$$I_d = I_S\left[\exp\left(\frac{v_D}{V_T}\right) - 1\right] + i_L. \qquad (9.20)$$

When the photodiode is in the photovoltaic mode, i.e. when its terminals are open, the photodiode is then typically used as a solar cell. The corresponding voltage at the diode open ends can be obtained from equation (9.17) for $i_D = 0$, i.e. generated voltage is:

$$v_D = V_T \,\mathrm{h}\left(1 + \frac{i_P}{I_S}\right). \qquad (9.21)$$

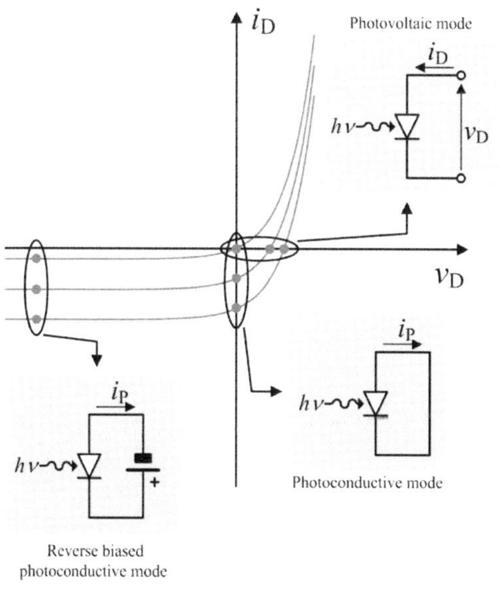

Figure 9.10. Photodiode operating regions.

Avalanche Photodiodes

Unlike the conventional photodiode, which can generate single electron-hole pair from a single absorbed photon, an avalanche photodiode (APD) can produce many electron-hole pairs from a single absorbed photon. The internal gain of the avalanche photodiode enables its use for constructing the highly sensitive optical receivers. The typical structure of an APD is presented in figure 9.11.(a).

As presented in figure 9.11.(a) the APD consists of four sections, which doping profile is presented in figure 9.11.(b). The leftmost section is a p-type semiconductor which is doped with the acceptor atoms with the concentration N_a and the rightmost section is a n-type semiconductor which is doped with the donor atoms with the concentration N_d. In between there is a middle section which is lightly doped semiconductor with a second p-type section. The corresponding charge density is presented in figure 9.11.(c), the energy diagram in figure 9.11.(d), and the electric field profile along the APD in figure 9.11.(e).

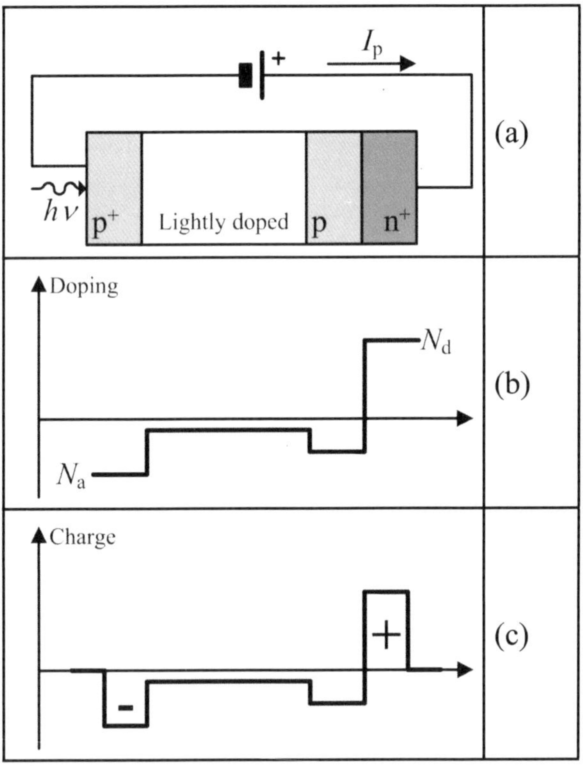

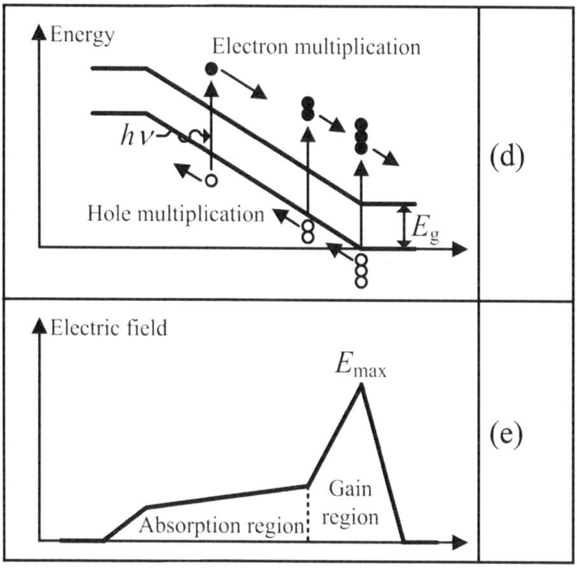

Figure 9.11. Avalanche photodiode as a photodetector.

In the case of high reverse biased p-n junction there are two breakdown mechanisms that can occur. The first mechanism implies that the atoms are directly ionized by the applied electric field. This is well known zener breakdown which is commonly used for making the voltage-regulating diodes. The second mechanism is caused by the high carrier velocity which causes the impact ionization within the semiconductor. This mechanism is called avalanche breakdown. The photons are absorbed within the lightly doped semiconductor material thus generating the electron-hole pairs. The generated carriers are quickly transported by the applied electric filed to the absorption region edges. After reaching the gain region the carriers are accelerated by the very high electric field thus achieving sufficient velocity (kinetic energy) to trigger the generation of additional electron-hole pair. This process is called the impact ionization. These additional carriers can then undergo other impact ionizations thus causing an avalanche effect. Therefore, an APD is similar to the photomultiplier where a single absorbed photon can generated a shower of the carriers. The corresponding current gain or multiplication factor of an APD is given by:

$$M = \frac{i_{APD}}{i_P}, \qquad (9.22)$$

where i_{APD} is the APD photocurrent and i_P is the internal photocurrent before the multiplication which is given by equation (9.19). The typical values of the APD gain ranges somewhere between few tens and few hundreds, although the gains as high as 10^3 or 10^4 can be reached.

REFERENCES

1. Giancarlo C. Righini, Antonella Tajani, and Antonello Cutolo. *An Introduction to Optoelectronic Sensors*. Singapore: World Scientific Publishing Co. Pte. Ltd., 2009.
2. Charles H. Cox, III. *Analog Optical Links: Theory and Practice*. Cambridge: Cambridge University Press, 2004.
3. Alan A. Stocker. *Analog VLSI Circuits for the Perception of Visual Motion*. Chichester: John Wiley & Sons Ltd, 2006.
4. Eduard Säckinger. *Broadband Circuits for Optical Fiber Communication*. Hoboken: John Wiley & Sons, Inc., 2003.
5. Carolien Hermans and Michiel Steyaert. *Broadband Opto-Electrical Receivers in Standard CMOS*. Dordrecht: Springer, 2007.
6. Mohamed Henini and Manijeh Razeghi (Eds.). *Handbook of Infrared Detection Technologies*. Oxford: Elsevier Science Ltd., 2002.
7. Kerstin Schneider and Horst Zimmermann. *Highly Sensitive Optical Receivers*. Berlin: Springer Berlin Heidelberg, 2006.
8. Jozef Piotrowski and Antoni Rogalski. *High-Operating-Temperature Infrared Photodetectors*. Bellingham: The Society of Photo-Optical Instrumentation Engineers, 2007.
9. Mark Johnson. *Photodetection and Measurement: Maximizing Performance in Optical Systems*. New York: The McGraw-Hill Companies Inc., 2003.
10. Joachim Piprek. *Semiconductor Optoelectronic Devices: Introduction to Physics and Simulation*. Amsterdam: Elsevier Science (USA), 2003.
11. Graham T. Reed and Andrew P. Knights. *Silicon Photonics: An Introduction*. Chichester: John Wiley & Sons Ltd, 2004.
12. Leonardo Rossi, Peter Fischer, Tilman Rohe, and Norbert Wermes.

Pixel Detectors: From Fundamentals to Applications. Berlin: Springer-Verlag Berlin Heidelberg, 2006.

13. Stephan B. Alexander. *Optical Communications Receiver Design*. Bellingham: SPIE – The International Society for Optical Engineering, 1997.
14. Jerald Graeme. *Photodiode Amplifiers: Op Amp Solutions*. New York: McGraw-Hill Companies Inc., 1996.

INDEX

M

N

P

Q

R

S

T

U

V

W

Z